Chemistry Research and Applications

Chemistry Research and Applications

The Future of Biorefineries
Waldemar Nyström (Editor)
2023. ISBN: 979-8-88697-524-6 (Hardcover)
2023. ISBN: 979-8-88697-528-4 (eBook)

Properties and Uses of Antimony
David J. Jenkins (Editor)
2022. ISBN: 979-8-88697-081-4 (Softcover)
2022. ISBN: 979-8-88697-088-3 (eBook)

The Science of Carbamates
Güllü Kaymak (Editor)
2022. ISBN: 978-1-68507-708-2 (Softcover)
2022. ISBN: 978-1-68507-872-0 (eBook)

Deep Eutectic Solvents: Properties, Applications and Toxicity
Carlos Eduardo de Araújo Padilha, PhD, Everaldo Silvino dos Santos, PhD, Francisco Canindé de Sousa Júnior, PhD, Nathália Saraiva Rios, PhD (Editors)
2022. ISBN: 978-1-68507-719-8 (Hardcover)
2022. ISBN: 978-1-68507-799-0 (eBook)

Polycyclic Aromatic Hydrocarbons: Sources, Exposure and Health Effects
Warren L. Gregoire (Editor)
2022. ISBN: 978-1-68507-626-9 (Softcover)
2022. ISBN: 978-1-68507-685-6 (eBook)

Cyanide: Occurrence, Applications and Toxicity
Bill M. Torres (Editor)
2022. ISBN: 978-1-68507-619-1 (Softcover)
2022. ISBN: 978-1-68507-670-2 (eBook)

More information about this series can be found at
https://novapublishers.com/product-category/series/chemistry-research-and-applications/

Scott R. Sheley
Editor

The Chemistry of Coumarin

Copyright © 2023 by Nova Science Publishers, Inc.

All rights reserved. No part of this book may be reproduced, stored in a retrieval system or transmitted in any form or by any means: electronic, electrostatic, magnetic, tape, mechanical photocopying, recording or otherwise without the written permission of the Publisher.

We have partnered with Copyright Clearance Center to make it easy for you to obtain permissions to reuse content from this publication. Simply navigate to this publication's page on Nova's website and locate the "Get Permission" button below the title description. This button is linked directly to the title's permission page on copyright.com. Alternatively, you can visit copyright.com and search by title, ISBN, or ISSN.

For further questions about using the service on copyright.com, please contact:
Copyright Clearance Center
Phone: +1-(978) 750-8400 Fax: +1-(978) 750-4470 E-mail: info@copyright.com.

NOTICE TO THE READER

The Publisher has taken reasonable care in the preparation of this book, but makes no expressed or implied warranty of any kind and assumes no responsibility for any errors or omissions. No liability is assumed for incidental or consequential damages in connection with or arising out of information contained in this book. The Publisher shall not be liable for any special, consequential, or exemplary damages resulting, in whole or in part, from the readers' use of, or reliance upon, this material. Any parts of this book based on government reports are so indicated and copyright is claimed for those parts to the extent applicable to compilations of such works.

Independent verification should be sought for any data, advice or recommendations contained in this book. In addition, no responsibility is assumed by the Publisher for any injury and/or damage to persons or property arising from any methods, products, instructions, ideas or otherwise contained in this publication.

This publication is designed to provide accurate and authoritative information with regard to the subject matter covered herein. It is sold with the clear understanding that the Publisher is not engaged in rendering legal or any other professional services. If legal or any other expert assistance is required, the services of a competent person should be sought. FROM A DECLARATION OF PARTICIPANTS JOINTLY ADOPTED BY A COMMITTEE OF THE AMERICAN BAR ASSOCIATION AND A COMMITTEE OF PUBLISHERS.

Additional color graphics may be available in the e-book version of this book.

Library of Congress Cataloging-in-Publication Data

ISBN: 979-8-88697-560-4

Published by Nova Science Publishers, Inc. † New York

Contents

Preface		vii
Chapter 1	**One-Pot Synthesis of Bromocoumarins via Consecutive Pechmann Condensation and Bromination Catalyzed by Cellulose Sulfuric Acid** Palani Natarajan	1
Chapter 2	**Recent Advances of Heterocylic Compounds** Sharanabasappa B. Patil	13
Chapter 3	**Chemical Constituents and Synthetic Methods of Bioactive Coumarins** Li Huang, Hui-Jing Li and Yan-Chao Wu	23
Chapter 4	**Insights into the Structure-Activity Relationship of Alkynyl-Coumarinyl Ethers as Selective MAO-B Inhibitors Using Molecular Docking** Yassir Boulaamane, Mohammed Reda Britel and Amal Maurady	77
Chapter 5	**Phytochemistry and Pharmacological Actions of Coumarin** Ramesh Vimalavathini and Gnanakumar Prakash Yoganandam	93
Chapter 6	**An Overview of Coumarins: Pharmacognosy, Phytochemistry and Structural Activity Relationship (SAR)** Gnanakumar Prakash Yoganandam, Ramesh Vimalavathini and Meenatchisundaram Sakthiganapathi	107
Index		121

Preface

This book reviews and examines the chemistry of coumarin from different perspectives in six chapters. Chapter One reviews the one-pot synthesis of bromocoumarins via consecutive Pechmann condensation and bromination catalyzed by cellulose sulfuric acid. Chapter Two examines recent advances in heterocyclic compounds. Chapter Three looks at the chemical constituents and synthetic methods of bioactive coumarins. Chapter Four provides insights into the structure-activity relationship of alkylnyl-coumarinyl ethers as selective MAO-B inhibitors using molecular docking. Chapter Five reviews the phytochemistry and pharmacological actions of coumarin. Chapter Six is an overview of coumarins from different perspectives.

Chapter 1 - More than 10 bromocoumarin derivatives have been synthesized using sequential two-step one-pot reactions with the help of a bio-supported, and widely accessible cellulose sulfuric acid as a catalyst. In step one, phenols and β-ketoesters undergo Pechmann condensation to produce coumarins, which are then reacted with potassium bromide to produce bromocoumarins. Additional appealing characteristics of this innovative methodology include straightforward starting materials, simpler work-up methods, and catalyst recovery and reuse at least five times.

Chapter 2 - Coumarin is a universal analog associated with various pharmacological potencies and has majorly contributed to the field of medicine. Coumarins are present naturally in high concentrations in Tonka bean, Vanilla grass, Cassia cinnamon, Justicia pectoralis, etc. A different type of coumarin exhibits various pharmacological potentials. In addition, coumarin coupled with pyrimidines, benzimidazoles, & other heteroatoms shows various medicinal significance such as anti (microbial, malarial, HIV, cancer, tubercular, thrombotic) antioxidant, & anticonvulsant activities. Based on the above biological potentials of the coumarin scaffold, the recent medicinal possibilities are discussed in this chapter.

Chapter 3 - Coumarins, with strong fluorescence and aromatic smell, are one of the main active ingredients in medicinal plants. Coumarins are present in free states or glycosides in plants, which are widely distributed in higher plants such as Umbelliferae, Rutaceae, Moraceae, Leguminosae, Oleaceae, and Compositae. Besides, a few coumarins are found in animals and microorganisms. Till now, about 1200 kinds of coumarins have been discovered. These coumarins are the general name of o-hydroxycinnamic acid lactones. The mother nucleus of coumarins is benzo α-pyranone. Other groups are hydroxyl, alkoxy, phenyl and isopentenyl substituents on the rings. The active double bond of isopentenyl and the ortho hydroxyl on the benzene ring can form the structure of furan ring or pyran ring. According to the different positions and characteristics of substituents in the structure of coumarins, they can be divided into simple coumarins, furancoumarins, pyranocoumarins and substituted α-pyranone coumarins. Moreover, coumarins exhibit obvious bioactivities, such as anti-tumor, anti-AIDS, anti-cell proliferation, anti-virus, anti-fungal, anti-bacterial, anti angiosclerosis, anti-oxidation, enhancing human immunity, etc. They are generally used in medicines. Additionally, coumarins show strong optical activity as fluorescent materials, including fluorescent brighteners, fluorescent dyes, laser dyes, organic photosensitive dyes for solar cells, fluorescent probes for biological analysis, etc. In recent years, new technologies, catalysts, and methods have been applied to the synthesis of coumarins. Coumarins are obtained by artificial synthesis or structural modification of coumarin parent nucleus to form new derivatives, and then further screening of the selective activities of these derivatives are evaluated. This has become a hot spot in the development and research of new coumarins as lead compounds. Briefly, coumarins reveal good bioactivities and a wide range of applications in medicines, foods, and materials, which are important and have high application value.

Chapter 4 - Coumarins are considered a highly privileged and versatile scaffold by medicinal chemists. A considerable number of studies have highlighted the synthesis and the various pharmacological activities of coumarins as promising drug candidates for treating neurodegenerative diseases such as Parkinson's and Alzheimer's disease. A wide range of compounds based on the coumarin ring system have been found to possess biological activities such as anticonvulsant, antiviral, anti-inflammatory, antibacterial, antioxidant as well as monoamine oxidase inhibitory properties. Their promise as a novel drug for neurodegenerative diseases is demonstrated by many drug candidates that made it to clinical trials such as nodakenin that have been potent for demoting memory impairment. This study focuses on some synthesized alkynyl-coumarinyl ethers with promising MAO-B inhibitory activity and selectivity and aims to elucidates the molecular interactions of ether-connected coumarins behind obtaining remarkably high MAO-B selectivity using molecular docking. Structure-activity relationship analysis revealed a common interaction between the selective coumarin inhibitors consisting of hydrogen bonding with Tyr-188 and Cys-172. The author's findings might open new opportunities to explore for developing novel highly selective MAO-B inhibitors for the treatment of neurodegenerative diseases.

Chapter 5 - Coumarins (2H-1-benzopyran-2-one) are the leading group of benzopyran derivatives that originate in plants. Coumarin name, is derived from a French term, Coumarou an aromatic, colorless compound, and was first isolated from the Tonka bean (Coumarouna odorata, Wild, family Fabaceae) in 1820. Coumarins are conveyed in about 150 diverse species distributed over nearly 30 different families, of which a few important are Apiaceae, Asteraceae, Rutaceae, Umbelliferae and Clusiaceae. Their structure consists of two six-membered rings with lactone carbonyl groups. Most coumarin compounds are thermally stable and have distinguished optical activity. They are biosynthesized from the phenyl propanoid pathway via ortho-hydroxylation. After hydroxylation, trans/cis isomerization and lactonization occur, resulting in the production of their respective coumarins. They have been the centre of attraction for medicinal chemist for the past few decades owing to their profound pharmacological activities like anti-microbial, anti-oxidant, anti-inflammatory, analgesic, anti-cancer, anti-malarial, anti-hyperlipidemic, anti-epileptic, anti-parkinsonian, anti-hepatitis, anti-coagulant, enzyme inhibiton and vasorelaxant properties. This chapter is a write up on the occurrence, phytochemistry and therapeutic actions of coumarins.

Chapter 6 - Coumarins (2H-1-benzopyran-2-one) are named based on the plant Coumarouna odorata (Dipteryx odorata, Family: Fabaceae), from which it was first isolated by Vogel in 1820. Coumarins are secondary metabolites existing in a wide array of higher plants and also in some microorganisms and animal species. In the plant kingdom, Coumarin occurs in both monocotyledonous and dicotyledonous plants and found in plant families such as Umbelliferae, Rutaceae, Compositae, Leguminosae, Oleaceae, Moraceae and Thymelaeacea. Coumarins are present in different plant body part including roots, leaves, flowers, fruits, roots and in the exudates of plants. The biosynthesis of Coumarin in plants occurs through hydroxylation, glycolysis and cyclization of Cinnamic acid. Dietary exposure to benzopyrones is significant as these compounds are found in vegetables, fruits, seeds, nuts, coffee, tea, and wine. More than 1300 Coumarins have been identified as secondary metabolites from plants, bacteria, and fungi. The prototypical compound is known as 1, 2-benzopyrone or, less commonly, as hydroxycinnamic acid and lactone, and it has been well studied. Although distributed throughout all parts of the plant, the Coumarins occur at the highest levels in the fruits, Bael fruits (Aegle marmelos), seeds, Tonka beans (Calophyllum inophyllum), roots (Ferulago campestris), leaves (Murraya paniculata) and latex of the tropical rainforest tree (Calophyllum teysmannii var. inophylloide). They are also found at high levels in some essential oils such as Cassia oil, Cinnamon bark oil, and Lavender oil. Environmental conditions and seasonal changes could influence the incidence of Coumarins in varied parts of the plant. Structure activity relationship (SAR) attempts to establish the suitability of various functional groups or moieties at different positions of a pharmacophore nucleus and is thus exploited for optimization of drug receptor interaction. SAR studies of coumarins reveal that its electronegativity and aromaticity impart plethora of pharmacological activity.

Chapter 1

One-Pot Synthesis of Bromocoumarins via Consecutive Pechmann Condensation and Bromination Catalyzed by Cellulose Sulfuric Acid

Palani Natarajan*

Department of Chemistry & Centre for Advanced Studies in Chemistry,
Panjab University, Chandigarh, India

Abstract

More than 10 bromocoumarin derivatives have been synthesized using sequential two-step one-pot reactions with the help of a bio-supported, and widely accessible cellulose sulfuric acid as a catalyst (Scheme). In step one, phenols and β-ketoesters undergo Pechmann condensation to produce coumarins, which are then reacted with potassium bromide to produce bromocoumarins (Scheme). Additional appealing characteristics of this innovative methodology include straightforward starting materials, simpler work-up methods, and catalyst recovery and reuse at least five times.

Scheme. The CSA catalyzed one-pot synthesis of bromocoumarins from phenols, β-ketoesters and potassium bromide.

---

* Corresponding Author's Email: pnataraj@pu.ac.in.

In: The Chemistry of Coumarin
Editor: Scott R. Sheley
ISBN: 979-8-88697-560-4
© 2023 Nova Science Publishers, Inc.

Keywords: metal-free, one-pot synthesis, bromocoumarins, Pechmann condensation

Introduction

Among aromatic lactones, coumarin (2H-1-benzopyran-2-one) [1] and its derivatives are the most well-known. Numerous microbes and plants, such as cinnamon, grass, lavender, protozoa, strawberries, and tonka bean, are responsible for their widespread distribution in nature [1, 2].

Particularly the brominated coumarins show biological characteristics including cytotoxicity, bioantioxidant, neurotropic, and renal cell carcinoma [3, 4]. Additionally, bromocoumarins have been used as precursors to create a variety of compounds, including the potentially useful mercapto-coumarins, and many others [5, 6]. Most often, combinations of bromine or bromide and sulfuric acid are used to brominate coumarins to produce bromocoumarins [7]. These techniques undoubtedly have some benefits, but they also have significant disadvantages, including lengthy work-up procedures, incompatibility with functional groups, over-bromination, toxic, expensive reagents, multi-step synthesis, high reaction temperatures, and inability to be recovered or reused.

In light of these limitations, efforts are still being made to develop a more effective and environmentally friendly approach for creating bromocoumarin derivatives.

Recent years have seen significant improvements in organic synthesis due to the introduction of solid acidic supports as catalysts [8-11]. Cellulose sulfuric acid (CSA, Scheme 1) has demonstrated enormous potentiality in this area.

It is also reusable, environmentally safe, and stable under a variety of reaction conditions. In addition, one of the most prevalent natural biopolymers in the world, cellulose is insoluble in ordinary organic solvents and is thought to be an endless source of resources to meet the growing demand for biocompatible materials [5].

This chapter describes a brand-new, one-pot procedure (Scheme 1) for the synthesis of bromocoumarins at room temperature that uses CSA (Scheme 1) as a catalyst and potassium bromide (Scheme 1) as a source of Br^+. To produce coumarins in situ, the reactions proceed through a Pechmann condensation reaction between activated phenols and -ketoesters, see Scheme 1. This technique uses straightforward catalysts and starting materials, produces moderate to good yields, and is metal-free.

Additionally, CSA is readily recoverable by filtration from the reaction mixture and can be recycled at least five times without noticeably losing its activity, see infra.

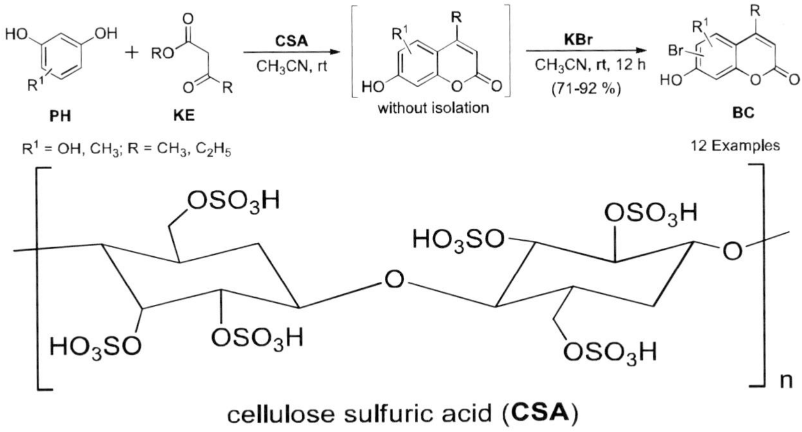

Scheme 1. In top is shown CSA catalyzed one-pot synthesis of bromocoumarins from phenols, β-ketoesters and KBr reported in this chapter. In bottom is dipicted the molecular structures of cellulose sulfuric acid (CSA).

Results and Discussion

In order to find the most appropriate reaction conditions and to evaluate the catalytic performance of **CSA** in Pechmann condensation reaction (Scheme 1); first, conducted a series of model studies involving resorcinol (**PH1**, 1mmol), ethyl acetoacetate (**KE1**, 1mmol) and **CSA** (0.1g) in different solvents such as CH_2Cl_2, $CHCl_3$, CH_3CN, toluene and 1,4-dioxane as well as under solvent-free conditions at room temperature. The results were collected in Table 1. Notice that: i) reactions worked well in the solution state than the solid state (Table 1 and Entries 1-5) and ii) CH_3CN to be a better solvent (yield 94%, Table 1 and Entry 3) than other solvents tested. Therefore, CH_3CN was chosen as the medium for all further reactions. To evaluate the effect of catalyst concentration, **PH1** (1mmol) and **KE1** (1mmol) was reacted in the presence of different amounts (0.025, 0.05, 0.075, 0.1 and 0.125g/1mmol of **PH1**) of **CAS** in CH_3CN at room temperature. Results indicated that 0.075 g of **CSA** was an optimum amount for effective condensation of phenols with β-ketoesters, cf. Table 1 and Entry 9. Use of catalyst in higher amounts (0.1 and 0.125g/1mmol of **PH1**) improves neither yield nor reaction time significantly. However, utilization of lower amounts (0.025 and 0.05g/1mmol of **PH1**) resulted to prolong reaction time with lower yield of expected coumarin. Interestingly, no reaction took place in the absence of the catalyst (Table 1 and Entry 12). Thus, the best results were obtained by carrying out the Pechmann condensation reaction with **PH** (1 mmol), **KE** (1 mmol) and **CSA** (0.075 g) in dry CH_3CN at room temperature, cf. Table 1.

Table 1. Selected results of screening the optimal conditions for Pechmann condensation reaction between phenols and β-ketoesters in the presence of CSA at room temperature[a]

Entry	Solvent [b]	CSA amount (g/1mmol of **PH1**)	Time (h)	Yield (%)
1	CH_2Cl_2	0.1	2	78
2	$CHCl_3$	0.1	2	81
3	CH_3CN	0.1	2	94
4	toluene	0.1	2	46
5	1,4-dioxane	0.1	2	10
6	neat	0.1	2	<5
7	CH_3CN	0.025	12	61
8	CH_3CN	0.05	6	74
9	**CH_3CN**	**0.075**	**2**	**95**
10	CH_3CN	0.1	2	94
11	CH_3CN	0.15	1.5	96
12	CH_3CN	0	12	NR

[a] Unless stated otherwise all reactions were performed with 1.0mmol of PH, 1.0mmol of KE and CSA in dry solvent at room temperature.
[b] Solvents were rigorously purified following the methods described in reference.
[c] Isolated yields.
NR: no reaction.

Since the acid catalysts are known to promote Pechmann condensation as well as *electrophilic* substitution reactions, thus integrated two-steps together in a sequential manner to synthesize bromocoumarins (**BC**) in a one-pot. Thus, the solution of **PH1** (1mmol), **KE1** (1mmol) and **CSA** (0.075g) in CH_3CN was stirred at room temperature for 2h; afterwards KBr (1.1mmol) was immediately added into this mixture and stirred for an additional 15h. The results of analysis of reaction mixture by thin layer chromatography and gas chromatography proved that about 89% of 6-bromocoumarin (**BC1**) was formed. Thus, the combinations of **PH** (1mmol), **KE** (1mmol), **CSA** (0.075g) and KBr (1.1mmol) in CH_3CN at room temperature are the optimized reaction conditions for a one-pot synthesis of **BC** as shown in Scheme 1.

After determining the optimum reaction conditions, the substrate scope for a one-pot two-step synthesis of bromocoumarins (Scheme 1) was investigated. At the start, the reactions between various activated phenols (**PH2-PH7**), **KE1** and KBr were looked and the results are summarized in Table 2. Notice that the reaction was significantly affected by the electronic effects as well as position of substituents on the aromatic ring. For instance, substrates (**PH1-PH3** and **PH5-PH6**) having more than one –OH groups gave maximum products in a short period of time than the simple phenol (**PH7**). In addition, a strong dependence on the position of the substituents was observed. For example, the reaction with *meta*-hydroxyl substituted phenols afforded higher yield than the catechol (**PH4**, Table 2). This is likely attributed to the steric as well as electronic factors of **PH4** as similar findings were well documented in the literatures. Next, varied the substituents on the β-ketoester. Among four β-ketoesters such as ethyl acetoacetate (**KE1**, Table 2 and Entries 1-6), methyl acetoacetate (**KE2**, Table 2 and Entries 7-11) and ethyl 2-methylacetoacetate (**KE3**, Table 2 and Entry 12) were tested, regardless of the substituents, the desired bromocoumarins were obtained in moderate to good yields from all substrates. This method thus provides bromocoumarins in one-pot under mild reaction conditions.

Further, investigated the reusability of **CSA**; as it is critical for large-scale synthesis in both academics and industries [8, 11]. Under the optimized conditions, a reaction of **PH3** (1.0mmol), **KE1** (1.0mmol), **KBr** (1.1mmol) and **CSA** (0.075g) in CH_3CN was taken as a model to explore this experiment. After the reaction, product was dissolved in excess of CH_3CN by sonication. Subsequently, the catalyst was recovered by filtration and washed with CH_3CN and reused as such for subsequent experiments. The results presented in Figure 1 shows that the catalyst can be used without appreciable loss of activity.

Table 2. CSA catalyzed one-pot synthesis of bromocoumarins from activated phenol derivatives, β-ketoesters and KBr at room temperature [a]

R^1 = OH, CH_3; R = CH_3, C_2H_5

Entry	Phenol derivative[b]	β-ketoester[b]	Product[c]	Time (h)	Yield[d] (%)
1	PH2	KE1	BC2	24	83
2	PH3	KE1	BC3	20	79

Entry	Phenol derivative[b]	β-ketoester[b]	Product[c]	Time (h)	Yield[d] (%)
3	![PH4: catechol with OH, OH] **PH4**	![KE1: methyl acetoacetate] **KE1**	--	36	NE
4	![PH5: pyrogallol] **PH5**	**KE1**	![BC5: 6-bromo-7,8-dihydroxy-4-methylcoumarin] **BC5**	20	84
5	![PH6: phloroglucinol] **PH6**	**KE1**	![BC6: 6,8-dibromo-5,7-dihydroxy-4-methylcoumarin] **BC6**	26	78
6	![phenol] **PH6**	**KE1**	![BC7: 6-bromo-4-methylcoumarin] **BC7**	36	<5

Table 2. (Continued)

Entry	Phenol derivative[b]	β-ketoester[b]	Product[c]	Time (h)	Yield[d] (%)
7	PH1 (resorcinol)	KE2 (methyl acetoacetate)	BC1	16	82
8	PH2 (5-methylresorcinol)	KE2	BC2	24	77
9	PH3 (2-methylresorcinol)	KE2	BC3	22	79
10	PH5 (pyrogallol)	KE2	BC5	20	83

Entry	Phenol derivative[b]	β-ketoester[b]	Product[c]	Time (h)	Yield[d] (%)
11	PH6 (3,5-dihydroxyphenol)	KE2 (methyl acetoacetate)	BC6	26	84
12	PH1 (resorcinol)	KE3 (ethyl 2-methylacetoacetate)	BC8	20	76

[a] Unless stated otherwise all reactions were performed with 1.0mmol of **PH**, 1.0mmol of **KE**, CSA (0.075g) and KBr (1.1mmol) source in CH_3CN at room temperature.

[b] phenol derivatives and β-ketoesters were purified prior to use.

[c] literature known compounds.

[d] Isolated yield was calculated based on phenol derivatives.

NE: no expected product detected.

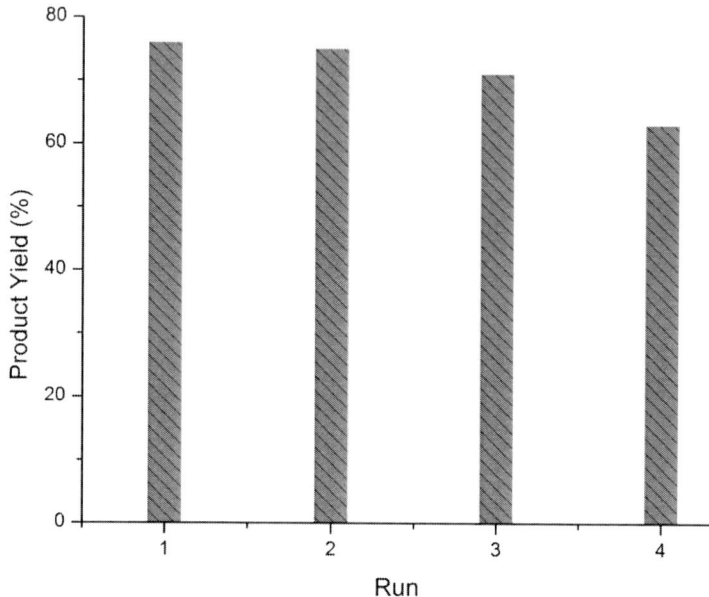

Figure 1. Recovery and reuse of CSA catalyst in the reaction of PH3 (1.0mmol), KE1 (1.0mmol), KBr (1.1mmol) and CSA (0.075g) in CH3CN under the optimized conditions.

Summary

The cellulose sulfuric acid as an efficient and environmentally friendly bio-supported catalyst provides proton source for the synthesis of bromocoumarins in a one-pot procedures. The simple experimental procedure, catalyst reusability, cost effectiveness, mild reaction conditions and moderate to good yields are salient features of this methodology.

Experimental

Synthesis

Materials and Methods
All solvents, precursors, reagents and catalyst were obtained from commercial suppliers and were dried or recrystallized before use. All

reactions were carried out in an inert atmosphere with standard Schlenk techniques. Reactions were monitored by analytical thin layer (0.06-0.15mm) chromatography on silica gel with analytical grade solvents as the eluents. The spots on TLC were visualized under UV light (λ = 254nm) or iodine chamber. Column chromatography was performed on silica gel or neutral alumina with 100-200 mesh. All NMR spectra were recorded on a Bruker Avance (300MHz) spectrometer in DMSO-d_6 or its mixture with CDCl$_3$: chemical shifts are expressed in parts per million (ppm) and were calibrated using the residual protonated solvent peak. IR spectra were recorded on a Perkin-Elmer IR spectrophotometer and stretching frequencies of only diagnostic bands such as carbonyl, hydroxyl, olefin, etc., are reported.

General procedure for the one-pot synthesis of bromocoumarins: To an oven-dried Schlenk tube equipped with a magnetic stir bar was charged with phenolic compound (1mmol), β-ketoester (1mmol), cellulose sulfuric acid (75mg) and dry CH$_3$CN. The suspension was stirred until the phenol was completely consumed (monitored by TLC; 1.5-2h). Instantly, KBr (1.1mmol) was added and the reaction mixture was allowed to stir further for the appropriate time (12-15h) at room temperature. Subsequently, the solvent was removed under reduced pressure. The resulting residue was dissolved in CH$_3$CN and insoluble catalyst was removed by filtration. The filtrate contains crude product was further purified by either recrystallization or filtration thru short pad silica gel/alumina column chromatography using either mixture of hexane-ethyl acetate or methanol-ethyl acetate.

References

[1] (a) Egan D, James P, Cooke D and O'Kennedy R. *Cancer Lett*, 1997, 118, 201; (b) Conley D and Marshall ME. *Proc Am Assoc Cancer Res*, 1987, 28, 63.
[2] Parfenov EA and Smirnov LD. *Khim Geterotsikl Soedi*, 1993, 4, 459.
[3] Savel'ev VL, Pryanishnikova NT, Artamonova OS, Fedina IV and Zagorevskii VA. *Khim-Farm Zh*, 1975, 9, 10.
[4] (a) Finn GJ, Kenealy E, Creaven BS and Egan DA. *Cancer Lett*, 2002, 183, 61; (b) Finn GJ, Creaven BS and Egan *DA. Cancer Letters*, 2004, 205, 69.
[5] (a) Majumdar KC and Ghosh SK. *J Chem Soc*, 1994, 2889; (b) Majumdar KC and Biswas P. *Tetrahedron*, 1999, 55, 1449.
[6] Parfenov EA, L. D. Smirnov LD. *Khim Geterotsikl Soedin*, 1990, 8, 1135.
[7] (a) Huebner CF and Link KP. *J Am Chem Soc*, 1945, 67, 99; (b) Cascaval A. *Synthesis*, 1985, 4, 428.

[8] (a) Alinezhad H, Haghigh A and Salehian F. *Chin Chem Lett*, 2010, 21, 183; (b) Sadaphal SA, Sonar SS and Ware MN. *Green Chem Lett Rev*, 2008, 1, 191.
[9] Shelke KF, Sapkal SB, Kakade GK, Bapurao B and Shingare MS. *Green Chem Lett Rev*, 2010, 3, 27.
[10] Oskooie HA, Heravi MM, Tahershamsi T, Sadjadi S and Tajbakhsh M. *Synth Commun*, 2010, 3, 1772.
[11] Shaabani A, Seyyedhamzeh M, Maleki A and Rezazadeh F. *Appl Catal A*, 2009, 358, 149.

Chapter 2

Recent Advances of Heterocylic Compounds

Sharanabasappa B. Patil*, PhD
Department of Chemistry, Ramaiah Institute of Technology, Bangalore, Karnataka, India

Abstract

Coumarin is a universal analog associated with various pharmacological potencies and has majorly contributed to the field of medicine. Coumarins are present naturally in high concentrations in Tonka bean, Vanilla grass, Cassia cinnamon, Justicia pectoralis, etc. A different type of coumarin exhibits various pharmacological potentials. In addition, coumarin coupled with pyrimidines, benzimidazoles, & other heteroatoms shows various medicinal significance such as anti (microbial, malarial, HIV, cancer, tubercular, thrombotic) antioxidant, & anticonvulsant activities. Based on the above biological potentials of the coumarin scaffold, the recent medicinal possibilities are discussed in this chapter.

Keywords: coumarin significance, synthetic approaches, medicinal applications

Introduction

Coumarin is a universal analog associated with various pharmacological potencies and a significant contribution to medicinal chemistry. Coumarin analogs act like anti (microbial, malarial, HIV, cancer, tubercular, thrombotic)

---

* Corresponding Author's Email: sbp7910@gmail.com.

In: The Chemistry of Coumarin
Editor: Scott R. Sheley
ISBN: 979-8-88697-560-4
© 2023 Nova Science Publishers, Inc.

antioxidant & anticonvulsant activities and are also found in medicinal plants and possess various biological activities (Figure 1) Based on the above therapeutic significance, this chapter further highlights coumarin's restorative capacities to prove a universal scaffold.

Literature Review

Classification of Coumarin

Coumarins are classified into Simple, Pyrano and Furo-coumarins. The different classes of coumarin have been associated with various biological potentials (Wu Y et al. 2020 & Javad Sharifi-Rad et al. 2021) (Figure 2).

Coumarin Drugs Are Commercially Available

The coumarin-associated drugs are commercially available to treat the various diseases (Warfarin, Acenocoumarol, Armillarisin A, & Auraptene), (Dinesh S Reddy et al. 2021) (Figure 3).

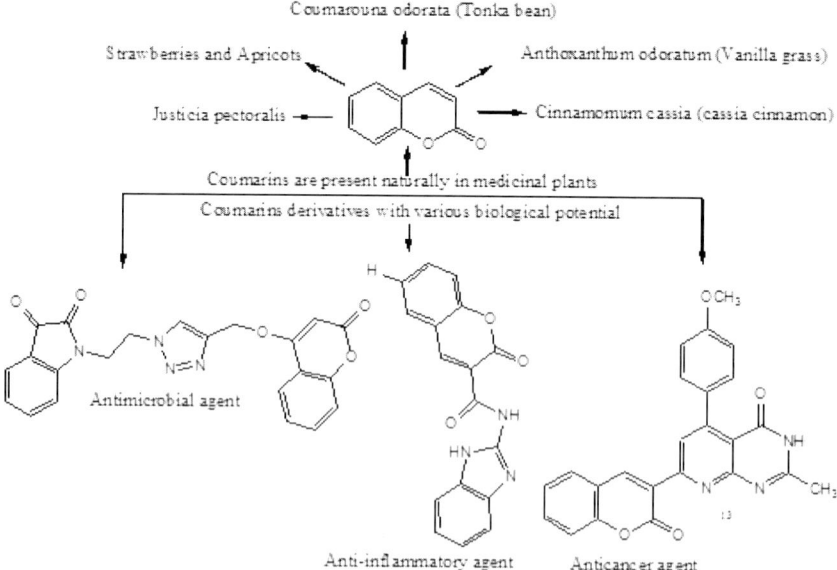

Figure 1. Coumarin presence in medicinal plants and Coumarin analogs with various biological activities.

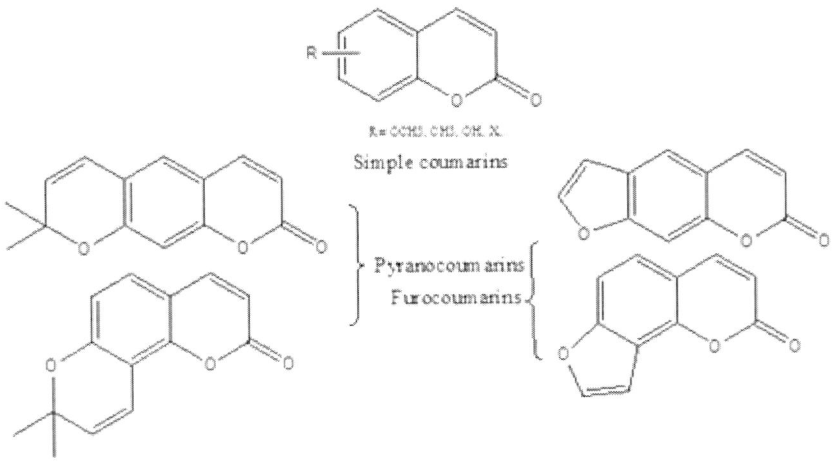

Figure 2. Coumarins structures of different class.

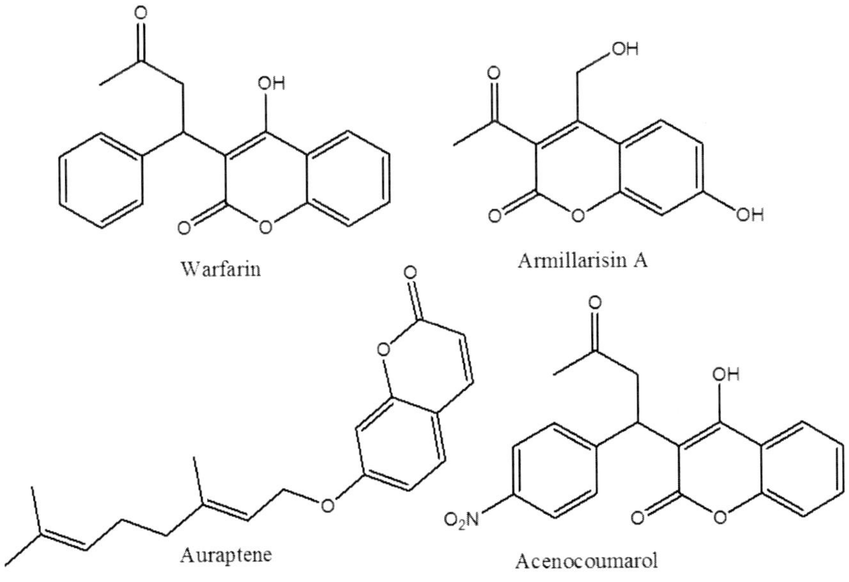

Figure 3. Clinically approved coumarin drugs.

Synthetic Approaches of Coumarins

Coumarin Synthesis from Aldehyde

Synthesis of coumarin derivatives by Knoevenagel condensation: Eutectic solvent was prepared (Just by blending choline chloride and zinc chloride at 100°C). Salicylaldehyde (substituted/Simple) & methylene compounds were used (Keshavarzipour F et al. 2016 & Loncaric M et al. 2020) (Figure 4).

Coumarin Synthesis from Phenols

Coumarin is synthesized by condensation of phenols with β-ketoesters & Sulfuric acid used as a catalyst (Resorcinol + ethyl acetoacetate + sulfuric acid at 80°C), (Rezaei R et al. 2014 & Loncaric M et al. 2020) (Figure 5).

Coumarin Synthesis from Ketones

Coumarin synthesized by condensation of 2-hydroxy-acetophenones & Meldrum's acid in green solvents at 60°C (Fiorito et al. 2016 & Loncaric M et al. 2020) (Figure 6).

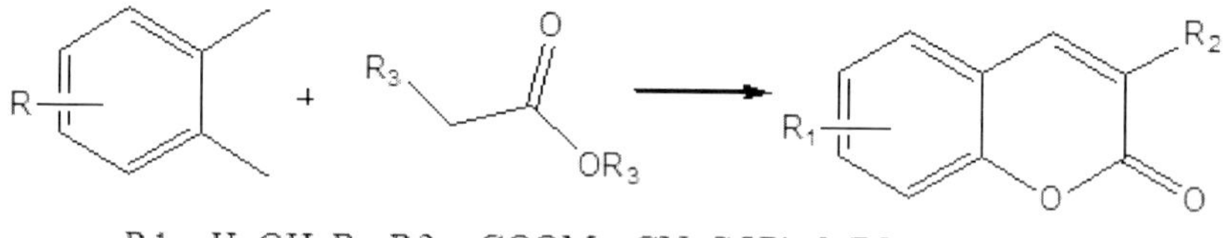

$R1 = H, OH, Br, R2 = COOMe, CN, COPh$ & $R3 = Me, Et$

Figure 4. Coumarin synthesis from aldehyde.

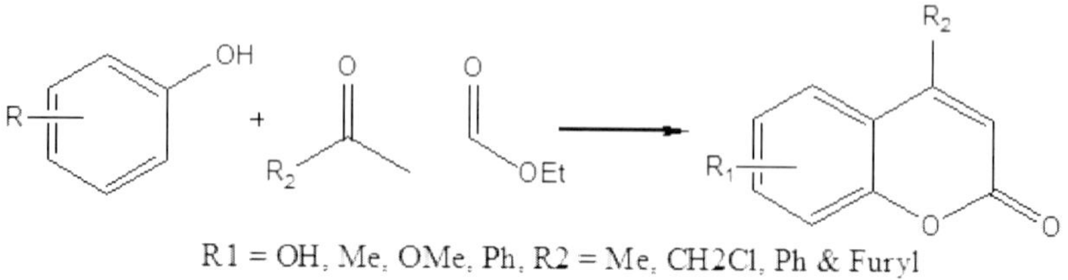

$R1 = OH, Me, OMe, Ph, R2 = Me, CH2Cl, Ph$ & $Furyl$

Figure 5. Coumarin synthesis from phenols.

Biological Potential of Coumarin Analogs

The biological potential of the coumarin scaffold is associated with various heteroatoms. It exhibits potent pharmacological activities such as antimicrobial, anticancer, anti-inflammatory, antimalarial, anti-HIV, anti-tubercular, antioxidant, & anticonvulsant agents (Table 1).

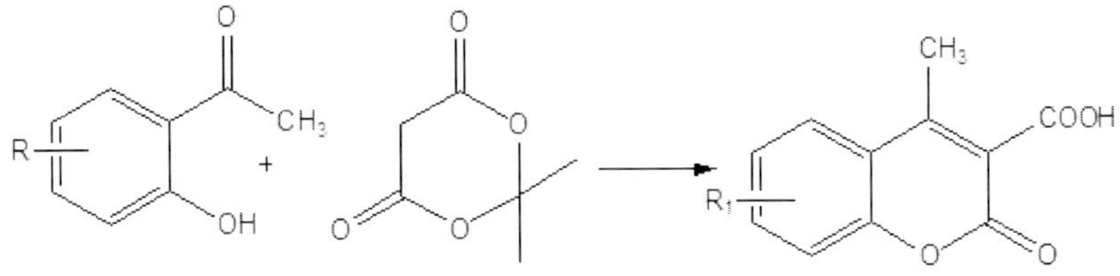

R1 = NO3, Cl, OMe & OH

Figure 6. Coumarin synthesis from ketones.

Table 1. Coumarin with various biological potential

1	as antimicrobial agent	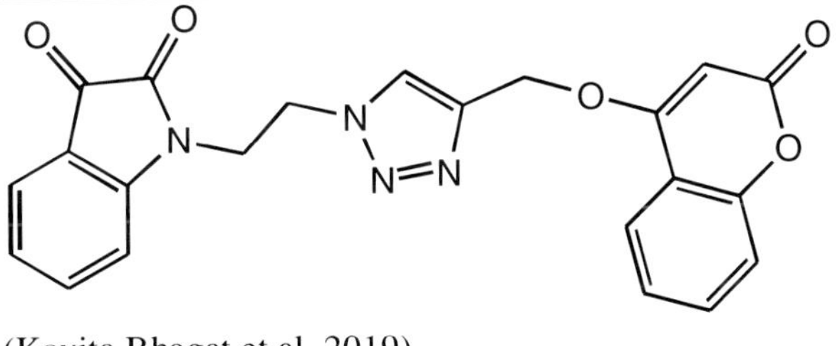 (Kavita Bhagat et al. 2019)
2	as anticancer agent	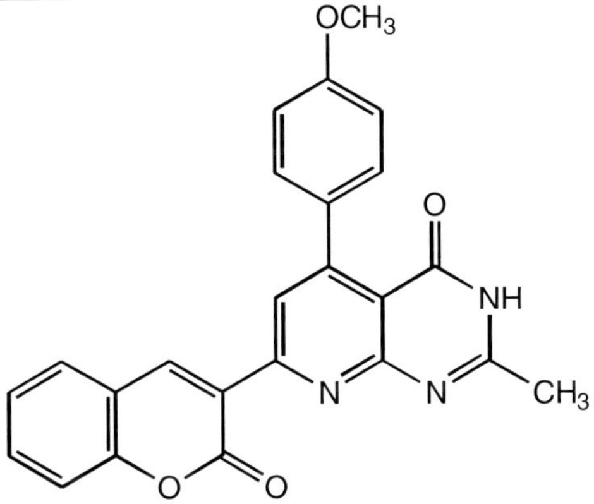 (Eman A. Fayed et al. 2019)

Table 1. (Continued)

3	as anti-inflammatory agent	(Radha Krishan Arora et al. 2014)
4	as anti-malarial agent	(Lorena Coronado et al. 2021)
5	as anti-HIV agent	(Yaseen A Al-Soud et al. 2007)

6	as anti-tubercular agent	(structure) (Godge R et al. 2018)
7	as antioxidant agent	(structure) (Donia Bensalah et al. 2020)
8	as anticonvalasant agent	(structure) (Ozan Tapanyigit et al. 2020)
9	as antiviral agent	(structure) (Rahman A et al. 2022)

Conclusion

Coumarins are associated with the benzopyrone family and are also found in medicinal plants in high concentrations. A broad variety of coumarin analogs play a key task in different biological pathways such as anti (microbial, malarial, HIV, cancer, tubercular, thrombotic) & antioxidant activities. In addition, coumarin analogs made a significant contribution to the medicinal field by developing a new drug. Recent articles on coumarins and other heteroatom reveal the broad spectrum of therapeutical applications (N. M. Goudgaon et al. 2009-2012, 2019 & S B Patil et al. 2018, 2020, 2021 & 2022).

References

Abdizadeh, R., Hadizadeh, F., & Abdizadeh, T. (2022). *In silico* analysis and identification of antiviral coumarin derivatives against 3-chymotrypsin-like main protease of the novel coronavirus SARS-CoV-2. *Mol. Divers* 26:1053-1076.doi.org/10.1007/s11030-021-10230-6.

Alshibl, Hanan M., Ebtehal S. Al-Abdullah, Mogedda E. Haiba, Hamad M. Alkahtani, Ghada E.A. Awad, Ahlam H. Mahmoud, Bassant M.M. Ibrahim, Ahmed Bari, and Alexander Villinger (2020). Synthesis and Evaluation of New Coumarin Derivatives as Antioxidant, Antimicrobial, and Anti-Inflammatory Agents. *Molecules*. 25(14): 3251. doi.org/10.3390/molecules25143251.

Al-Soud, Y. A., Sa'doni, H. H., Amajaour, H. A. S., Salih, K. S. M., Mubarak, M. S., Al-Masoudi, N. A., & Jaber, I. H. (2007). Synthesis, Characterization and anti-HIV and Antitumor Activities of New Coumarin Derivatives. *Z. Naturforsch*. 63b: 83-89.

Arora, R. K., Kaur, N., Bansal, Y., & Bansal, G. (2014). Novel coumarin-benzimidazole derivatives as antioxidants and safer anti-inflammatory agents. *Acta Pharmaceutica Sinica B*. 4(5): 368-375.doi.org/10.1016/j.apsb.2014.07.001.

Bensalah, Donia, Aziza Mnasri, A. Chakchouk-Mtibaa, Lamjed Mansour, L. Mellouli, and Naceur Hamdi (2020). Synthesis and antioxidant properties of some new thiazolyl coumarin derivatives, *Green Chemistry Letters and Reviews*, 13:2, 155-163. doi: 10.1080/17518253.2020.1762935.

Coronado, L., Zhang, X. Q., Dorta, D., Escala, N., Pineda, L. M., Ng, M. G., del Olmo, E., Wang, C. Y., Gu, Y. C., Shao, C. L., & Spadafora, C. (2021). Semisynthesis, Antiplasmodial Activity, and Mechanism of Action Studies of Isocoumarin Derivatives. *J. Nat. Prod*. 84 (5): 1434-1441. doi.org/10.1021/acs.jnatprod.0c01032.

Fayed, Eman A., Rehab Sabour, Marwa F. Harras, and Ahmed B. M. Mehany (2019). Design, synthesis, biological evaluation and molecular modeling of new coumarin derivatives as potent anticancer agents. *Medicinal Chemistry Research*. doi:10.1007/s00044-019-02373-x.

Fiorito, Serena, Vito Alessandro Taddeo, Salvatore Genovese, and Francesco Epifano (2016). A green chemical synthesis of coumarin-3-carboxylic and cinnamic acids using crop-derived products and waste waters as solvents. *Tetrahedron Lett.* 57:4795-4798. doi: 10.1016/j.tetlet.2016.09.023.

Gulati, H. K., Bhagat, J., Singh, H., & Bedi, P. M. S. (2019). Design, Synthesis, Antimicrobial Evaluation, and Molecular Modeling Studies of Novel Indolinedione-Coumarin Molecular Hybrids. *ACS Omega.* 4: 8720-8730. doi: 10.1021/acsomega.8b02481.

Godge, Rahul, and Rahul Kunkulol (2018). Synthesis of Coumarin heterocyclic derivatives with In-Vitro antituberculer activity, *Journal of Drug Delivery and Therapeutics.* 8(5):217-223. doi.org/10.22270/jddt.v8i5.1859.

Keshavarzipour, F., & Tavakol, H. (2016). The synthesis of coumarin derivatives using choline chloride/zinc chloride as a deep eutectic solvent. *J. Iran. Chem. Soc.* 13:149-153. doi: 10.1007/s13738-015-0722-9.

Lončarić, M., Gašo-Sokač, D., Jokić, S., & Molnar, M. (2020). Recent Advances in the Synthesis of Coumarin Derivatives from Different Starting Materials. *Biomolecules.* 16;10(1):151. doi: 10.3390/biom10010151.

N. M. Goudgaon et al. (2011). CNS depressant activity of some novel 5-substituted pyrimidin-2,4,6-Triones. *Journal of Pharmacy Research.* 4 (7):2195-2196.

N. M. Goudgaon et al. (2009). Synthesis and Antimicrobial evaluation of 5-iodopyrimidine analogs. *Indian J. Pharm. Sci.* 71 2009 672 10.4103/0250-474X.59551.

N. M. Goudgaon, et al. (2009). Synthesis of 2-benzylthiopyrimidinyl Pyrazole analogs and their antimicrobial activities, *Indian J. Heterocycl. Chem.* 18: 349-352.

N. M. Goudgaon, et al. (2009). Synthesis and Anti-microbial activity of Thiazole substituted Coumarins, *Heterocyclic Commu.* 15 (5) 343-348. doi. org/10.1515/HC.2009.15.5.343.

N. M. Goudgaon, et al. (2019). Synthesis and antimicrobial activity of novel coumarone analogues, *Int. J. Pharm. Sci. Res. IJPSR.* 10(2): 960-965. doi.org/10.13040/IJPSR.0975-8232.

N. M. Goudgaon, et al. (2010). Synthesis and antimicrobial activities of novel 5-substituted pyrimidin-2,4,6-triones, *J. Indian Chem. Soc.* 87: 743-748.

N. M. Goudgaon, et al. (2012). A facile route for the synthesis of novel 2-benzylthio-4,6-disubstituted pyrimidine analogues, *Ind. J. Heterocycl. Chem.* 21: 221-224.

Reddy, Dinesh S., Manasa Kongot, and Amit Kumar (2021). Coumarin hybrid derivatives as promising leads to treat tuberculosis: Recent developments and critical aspects of structural design to exhibit anti-tubercular activity. *Tuberculosis.* 127: 102050. doi.org/10.1016/j.tube.2020.102050.

Rezaei R et al. (2014). Coumarin synthesis via Pechmann condensation utilizing starch sulfuric acid as a green and efficient catalyst under solvent-free conditions. *Org. Chem. Indian J.* 2014;10:73-78.

S. B. Patil (2018). Biological and Medicinal Significance of Pyrimidine's: A Review, *International Journal of Pharmaceutical Sciences and Research.* 9(1): 44-52. doi: 10.13040/IJPSR.0975-8232.

Sharanabasappa B. Patil (2020). Biological and Pharmacological Significance of Benzimidazole Derivatives: A Review. *IJPSR,* 11(6): 2649-2654. doi: 10.13040/IJPSR.0975-8232. 11(6).2649-54.

Sharanabasappa B. Patil et al. (2021). Medicinal Significance of Coumarin Anaoogues: A Review. *Int. J. Curr. Pharm. Res.* 13(4):1-5. doi:10.22159/ijcpr.2021v13i4.42733.

Sharanabasappa B. Patil (2022). Medicinal significance of novel coumarin analogs: Recent Studies. *Results in Chemistry*. 100313. doi.org/10.1016/j.rechem.2022.100313.

Sharifi-Rad, J., Cruz-Martins, N., López-Jornet, P., Lopez, E. P. F., Harun, N., Yeskaliyeva, B., Beyatli, A., Sytar, O., Shaheen, S., Sharopov, F., Taheri, Y., Docea, A. O., Calina, D., & Cho, W. C. (2021). Review Article Natural Coumarins: Exploring the Pharmacological Complexity and Underlying Molecular Mechanisms. *Hindawi Oxidative Medicine and Cellular Longevity*. 6492346, 1-19. doi.org/10.1155/2021/649234.

Tapanyiğit, Ozan, Onur Demirkol, Ece Güler, Mehmet Erşatır, Muhammet Emin Çam, and Elife Sultan Giray (2020). Synthesis and investigation of anti-inflammatory and anticonvulsant activities of novel coumarin-diacylated hydrazide derivatives. *Arabian Journal of Chemistry*. 13 (12): 9105-9117. doi.org/10.1016/j.arabjc.2020.10.034.

Wu, Y., Xu, J., Liu, Y., Zeng, Y., & Wu, G. (2020). A Review on Anti-Tumor Mechanisms of Coumarins. *Front. Oncol.* 10:592853. doi: 10.3389/fonc.2020.592853.

Chapter 3

Chemical Constituents and Synthetic Methods of Bioactive Coumarins

Li Huang, PhD
Hui-Jing Li, PhD
and Yan-Chao Wu*, PhD

Weihai Key Laboratory of Active Factor of Marine Products, Weihai Marine Organism & Medical Technology Research Institute, Harbin Institute of Technology, Weihai, P. R. China

Abstract

Coumarins, with strong fluorescence and aromatic smell, are one of the main active ingredients in medicinal plants. Coumarins are present in free states or glycosides in plants, which are widely distributed in higher plants such as Umbelliferae, Rutaceae, Moraceae, Leguminosae, Oleaceae, and Compositae. Besides, a few coumarins are found in animals and microorganisms. Till now, about 1200 kinds of coumarins have been discovered. These coumarins are the general name of o-hydroxycinnamic acid lactones. The mother nucleus of coumarins is benzo α-pyranone. Other groups are hydroxyl, alkoxy, phenyl and isopentenyl substituents on the rings. The active double bond of isopentenyl and the ortho hydroxyl on the benzene ring can form the structure of furan ring or pyran ring. According to the different positions and characteristics of substituents in the structure of coumarins, they can be divided into simple coumarins, furancoumarins, pyranocoumarins and substituted α-pyranone coumarins. Moreover, coumarins exhibit obvious bioactivities, such as anti-tumor, anti-AIDS, anti-cell proliferation, anti-

---

* Corresponding Author's Email: yanchaowu@hit.edu.cn.

In: The Chemistry of Coumarin
Editor: Scott R. Sheley
ISBN: 979-8-88697-560-4
© 2023 Nova Science Publishers, Inc.

virus, anti-fungal, anti-bacterial, anti angiosclerosis, anti-oxidation, enhancing human immunity, etc. They are generally used in medicines. Additionally, coumarins show strong optical activity as fluorescent materials, including fluorescent brighteners, fluorescent dyes, laser dyes, organic photosensitive dyes for solar cells, fluorescent probes for biological analysis, etc. In recent years, new technologies, catalysts, and methods have been applied to the synthesis of coumarins. Coumarins are obtained by artificial synthesis or structural modification of coumarin parent nucleus to form new derivatives, and then further screening of the selective activities of these derivatives are evaluated. This has become a hot spot in the development and research of new coumarins as lead compounds. Briefly, coumarins reveal good bioactivities and a wide range of applications in medicines, foods, and materials, which are important and have high application value.

Keywords: coumarins, application, synthesis, bioactivities

Abbrevations

3T3:	Mouse fibroblasts
5-FU:	5-fluorouracil
A549:	Human lung adenocarcinoma cell line
AMV RT:	Alfalfa mosaic virus
BCL2:	B-cell lymphoma-2
CDC25:	Cell division cycle factor 25
CDK-6:	Mitotic protein kinase 6
COS-7:	SV40 transformed kidney cells of African green monkey
Cys:	Cysteine
DNA:	Deoxyribonucleic acid
DU145:	Human prostate cancer cells
GC:	Gas chromatography
GSH:	Glutathione
H1N1:	Influenza virus A
HCT-116:	Human colon cancer cell 116
HCT-28:	Human colon cancer cell 28
Hcy:	Homocysteine
HEK293:	Human embryonic kidney 293 cells
HepG2:	Human hepatoma cells
HIV-1:	Acquired immune deficiency syndrome of type 1

HPLC:	High performance liquid chromatography
HT-29:	Human colon cancer cell
HUVEC:	Human umbilical vein endothelial cell
IC50:	Half maximal inhibitory concentration
K562:	Human chronic myeloid leukemia cell
MCF-7:	Human breast cancer cells
MDR KB VIN:	Multidrug resistance
MS:	Mass spectrum
OVCAR-3:	Human ovarian cancer cells
PANC-1:	Pancreatic cancer cells
RD:	Human malignant embryonic rhabdomyoma cells
RNA:	Ribonucleic acid
SW48:	Human colon adenocarcinoma cells

1. Introduction

1.1. Sources of Coumarins

In 1820, the first natural coumarin was isolated by Vogel from tonka beans of the family Fabaceae in Guyana. The name "coumarin" is from the French term "coumarou". Then, coumarins as natural organic compounds with aromatic smell are widely found in secondary metabolites of plants, animals, and microorganisms with the form of free coumarin glycosides or polymers. They are the most abundant benzoa-α-pyranone derivatives discovered (Aydin et al., 2022). Coumarins, natural secondary metabolites of flavonoids with 1,2-benzopyran ring as the basic parent skeleton, are the oxygenated heterocyclic compounds. Coumarins are the most abundant in Rutaceae, Moraceae and Umbelliffrae in these dicotyledonous plants, and more in Orchidaceae, Compositae, Leguminosae, Solanaceae and Oleaceae (Azagarsamy et al., 2014). Due to their structural diversities, most of the natural coumarins show oxygen-containing substituents at the C-7 position, and the first one isolated from umbelliferae is 7-hydroxycoumarins, namely umbelliferolactone. Therefore, both coumarin and umbelliferone can be used as the parent of other high-oxygen, allylated, geranium, farlactone and more complex coumarin derivatives, as shown in Figure 1. Till now, nearly 1200 coumarin compounds have been isolated from nature (Babaei et al., 2022).

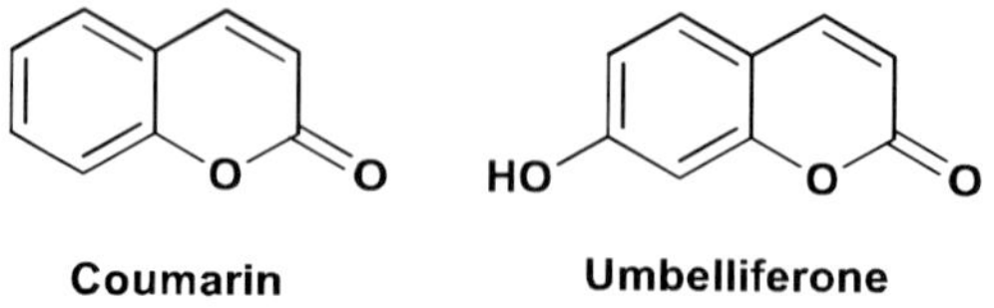

Figure 1. Structures of coumarin and umbelliferone.

1.2. Application and Progress of Coumarins

The exploration and application of coumarins in medicines, dyes and nonlinear optical materials have been widely concerned, thereby showing broad application prospects (Bonelli et al., 2022). Coumarins and derivatives reveal a variety of bioactivities, such as anti-HIV, antioxidant, anti-inflammatory, hepatoprotective, anti-thrombotic, antiviral, antibacterial, anti-tuberculosis, anti-depression, anti-hyperlipidemia, and anti-cancer (Cao et al., 2022; Cheke et al., 2022; Cho et al., 2022). A coumarin derivative, 8-MOP, is clinically used in the treatment of advanced cutaneous T-cell lymphoma and immune-related diseases. It is often used with ultraviolet light to activate 8-MOP, which increases the amount of reactive oxygen species released by the body and causes oxidative damage to DNA. Thus, cell proliferation and differentiation are greatly inhibited (Choi et al., 2022). Additionally, there are a variety of drugs with coumarins as precursors, such as scopolamine acting on the nervous system (Farhat et al., 2022), hydroxymethyl coumarin as an anticoagulant, and warfarin as an anticoagulant (Ren et al., 2018). Coumarins and their derivatives have been the preferred skeleton molecules in drug design.

Coumarins have a strong fragrance which are commonly used in perfumes, soaps, toilet water and beverages, and fragrance (Ferreira et al., 2022). Coumarins are also used as odor stabilizers in tobacco and masking agents in paint and rubber. Moreover, coumarins can be applied as food additives to improve the taste and quality of food (Franchin et al., 2022). For example, 3,4-dihydrocoumarin as an improver in perfume, and 6-methyl coumarin as a flavoring agent in ice cream (Freitas et al., 2022). Furthermore, coumarins are also used in pesticides, allelopathic agents, cosmetics, perfume fixatives, essential oils, and enhancers, demonstrating a great range of application value (Fu et al., 2022; Gao et al., 2022; Ghazouani et al., 2022; Hu et al., 2022).

Coumarin products with aromatic smell are often used as deodorants because they can mask the smell of phenols, iodoform, etc. (Houshmand et al., 2022; Liu et al., 2022a; Yang et al., 2022b). Coumarins usually give people a refreshing feeling, which greatly increase additional functions, and attract people's attention. More interestingly, coumarin compounds that perform well in agriculture are as pesticides. For example, osthole can effectively kill the young larvae of plutella xylostella and rapeseed (Xu et al., 2022). It is a promising direction to search for coumarin compounds that have effects on agricultural pests and pathogens, and then to develop biopesticides with low toxicity, high efficiency, and less residue.

Coumarins have been widely used as chemical sensors and luminescent materials due to their good photophysical properties. In most cases, coumarins are attached to recognition units as fluorescent molecules (Onder et al., 2022). In the past few years, coumarins were emerged as powerful and adaptable moleculars, but their photoself-polymerization under continuous light condition prevents their use as catalysts. Therefore, coumarins are introduced into supramolecular assembly and prevented from self-polymerization by steric hindrance or electrostatic action, which provides novel insights for the catalysis and recognition of coumarins with excellent photophysical properties (Naik et al., 2022). Notably, coumarins are most used as active ingredients in fluorescent brighteners, disperse fluorescent dyes, and laser dyes (Li et al., 2022b; Zhou et al., 2022a). Coumarins are widely applied in fluorescent dyes and fluorescent probes due to their high fluorescence intensity, good solubility, high quantum yield and convenient preparation. For example, calixarene derivatives are modified by fluorescent groups of coumarins to make them functional aromatic fluorescence sensors (Moghadam et al., 2022). Excitedly, the application of new coumarins is also gradually developed to organic light-emitting diode materials and solar cells.

2. Structural Types of Coumarins

There are two classification ways for coumarins. One is classified according to the position and number of oxysubstituents in coumarins (Jana et al., 2022). The other can be divided into simple coumarins, furancoumarins, pyrancoumarins and other coumarins according to the basic skeleton formed by biogenic pathway (Zhang et al., 2022a). Generally, the latter way is the main one.

2.1. Simple Coumarins

Simple coumarins are limited to the compounds with functional groups on the benzene ring. Briefly, there are most oxygen-containing groups at the C-7 position, and the common functional groups are -OH, -OCH$_3$, -OCHCH = C(CH$_3$)$_2$, etc. (Song et al., 2022). There are more isopentenes at the C-6 and C-8 positions which can be connected to the carbon chain or the oxygen atoms. Representative simple coumarins include esculetin, limettin, scopoletin, osthole, etc. (Figure 2).

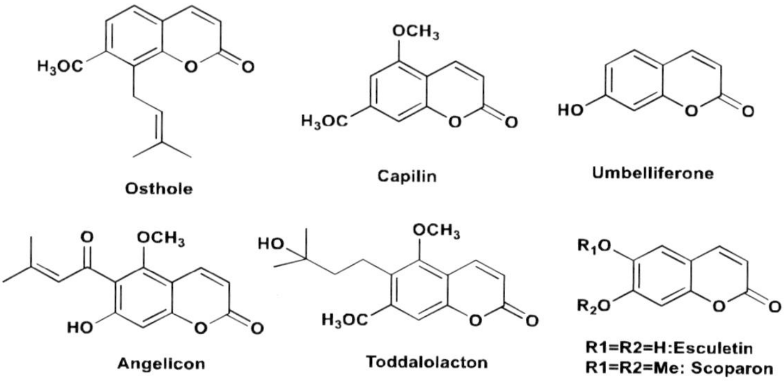

Figure 2. The structures of simple coumarins.

2.2. Furancoumarins

Furancoumarin is a ring of furan or dihydrofuran bound to the coumarin skeleton (Tong et al., 2022). The two rings are thickened in different ways to form various angular and linear furancoumarins. They are simple coumarin intermediates with isoprene structures which are produced by the coupling of umbelline lactone and dimethyl allyl pyrophosphate and then synthesized by phenylpropanoid and methylhydroxyvaleric acid, respectively. Linear furancoumarins are formed by the closed-loop reaction between the phenol hydroxyl group at the C-7 position and the isopentenyl group at the C-6 position, while angular furancoumarins are lactone compounds formed by the cyclization of the isopentenyl group at the C-8 position and the ortho phenol hydroxyl group (Ibrahim et al., 2022a; Mohammed et al., 2022). As described in Figure 3, the representative furancoumarins involve isober gapten and angelicin, and common linear furancoumarins include psoralen and xanthotol.

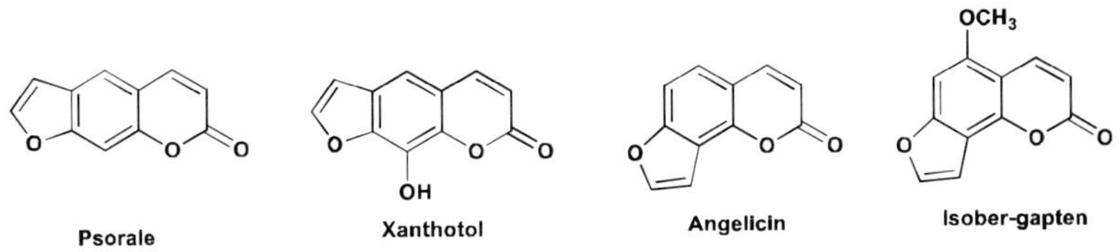

Figure 3. The structures of furancoumarins.

2.3. Pyranocoumarins

Pyranocoumarin is a pyran or dihydropyran ring formed by the condensation of isopentenyl at C-6 or C-8 position of 7-hydroxycoumarin and phenolic hydroxyl at C-7 position (Karcz et al., 2022). The closed-loop ways of the pyran ring led to the formation of different angular or linear pyranocoumarins. Pyranocoumarins and furancoumarins are basically the same in the biosynthetic pathway. The only difference between them is the ring formation from isoprene group to pyran ring, not furan ring. Xanthyletin, seselin and braylin are representative pyranocoumarins as Figure 4.

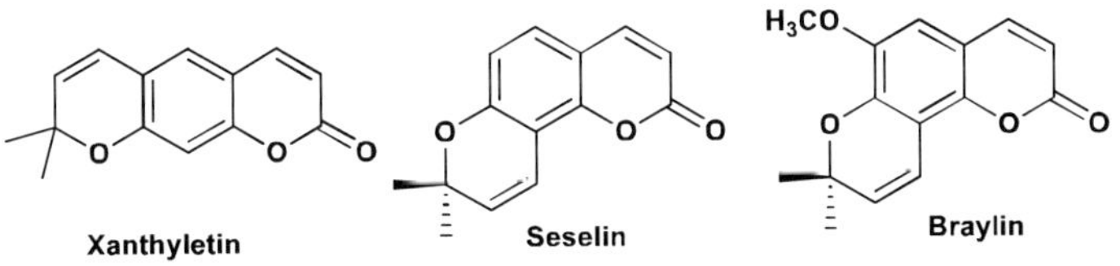

Figure 4. The structures of pyranocoumarins.

2.4. Other Coumarins

Other coumarins refer to these with substituents attached to the C-3 and C-4 positions of benzo-α-pyranone ring and coumarin polymers (Zhang et al., 2022b). In addition to C-3 and C-4 positions, there are also 3,4-benzo structures, such as glycyrrhizin, autumnariniol, wedelolactone, coumestrol, and hyuganin A. Currently, more than 1200 natural coumarins are isolated from plants, some of which show new sources and bioactivities (Long et al., 2022; Wang et al., 2022). New coumarins are also discovered, including coumarin dimers, isoprenodienylated furancoumarins, sesquiterpenyl coumarins, and some coumarins with special structures (Zhou et al., 2022b). Among them, isocoumarins exhibit good application values. They are isomers of coumarins whose chemical name is 1-*H*-2-benzopyran-1-ones, whose sources are like coumarins. In recent years, it has been found that isocoumarins and their derivatives also show antibacterial, anticancer, antiviral, anti-inflammatory, protease inhibitor, herbicidal and other biological and physiological activities in medicines and pesticides (Rani et al., 2022; Russell et al., 2022; Yu et al., 2022b). Their structures are shown in Figure 5.

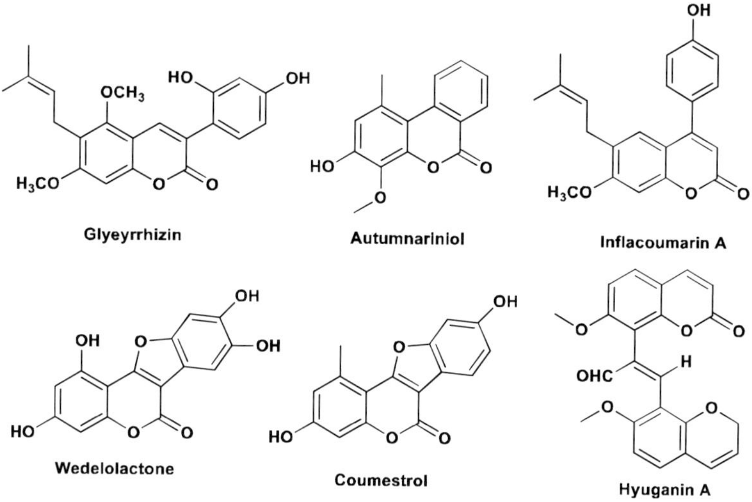

Figure 5. The structures of isocoumarins.

3. Extraction and Separation of Coumarins

3.1. Extraction Ways

3.1.1. Solvent Extraction

Most free coumarins are of low polarity. They are generally extracted with petroleum acid, benzene, acetic acid, ethyl acetate, acetone, and methanol in turn. Due to the high toxicity of benzene, it should be used less or not as much as possible (Kusumoto et al., 2022). Free coumarins are generally of low polarity. They can be directly crystallized or mixed crystallized after being extracted and concentrated with low polarity solvent and placed in a refrigerator. Sometimes the residue is too thick for the crystals to separate out, so a small amount of solvent can be added or the residue can be concentrated to the small amount and placed in the refrigerator for crystallization (Villa-Martinez et al., 2022). Besides, coumarins display higher polarity after they are combined with sugar to get glycosides. Hence, they can generally be extracted with a variety of solvents according to different polarity.

3.1.2. Acid-Base Method

Coumarins are separated since the lactone of coumarins can be recovered after opening the ring with alkali and acid. The ether solution containing coumarins is usually extracted with sodium bicarbonate aqueous solution to remove the acidic components, and then hydrolyzed with dilute and cold sodium hydroxide, potassium hydroxide aqueous solution or alcohol solution (Patil et al., 2022). Coumarins are just dissolved in water after being salted, and the neutral component that does not hydrolyze is extracted with ethanol. After neutralizing the remaining alkali solution with acid, the lactone components of coumarins are extracted with ether (Parvej et al., 2022). Finally, the acid obtained after saponification with other esters remains in the aqueous solution.

3.1.3. Steam Distillation

As for the volatile coumarins, the volatile oil can be extracted by steam distillation or pressing. Then the volatile oil is removed by steam distillation, and the total coumarins can be obtained by crystallization and recrystallization with organic solvents (Malankar et al., 2022).

3.1.4. Supercritical Fluid Extraction Technology

Supercritical fluid extraction technology has been widely used in the extraction of coumarins. Free coumarins with low polarity can be extracted directly, while glycosides can be prepared by entrainment with polar solvents such as ethanol.

3.2. Separation Ways

3.2.1. Crystallization

The crystallization of total coumarins can be precipitated from the concentrated solution of non-polar organic solvents after cooling. The leaching solution containing coumarins is extracted with mixed solvents. After concentration without crystallization, all the solvent can be recovered, and then obtained compounds are dissolved in petroleum ether and removed insoluble matter, and total coumarins are precipitated (Liu et al., 2022b).

3.2.2. Extraction Method

In the non-polar organic leaching solution of plant medicinal materials, sometimes there are acidic substances which can be extracted by dilute sodium bicarbonate aqueous solution. If coumarins contain hydroxyl, they can sequentially be extracted with dilute sodium hydroxide aqueous solution, and then adding acid to the aqueous solution to adjust pH to weak acidity, and finally phenol hydroxy coumarin can be precipitated.

3.2.3. Acid-Base Treatment

The organic solvent extracts which cannot precipitate the total coumarins are dissolved by adding dilute sodium hydroxide aqueous solution with slight heating to filter out insoluble substances. Then dilute acid is added to the filtrate to adjust the pH to weak acidity, and finally total coumarins are separated out (Li et al., 2022c). When using this method to refine coumarins, attention should be paid to prevent damage to alkali-sensitive substances.

3.2.4. Separation

As for the organic solvent containing coumarins and organic acids, these organic acids are extracted with dilute sodium bicarbonate aqueous solution, and then the phenolic components are extracted with dilute sodium hydroxide aqueous solution (Sartiva et al., 2022). Hydroxy coumarins are also obtained from the precipitate of sodium hydroxide extract. The neutral components of the organic solution are saponified in the ethanol solution of potassium hydroxide at room temperature. When the saponification is completed, then water is added and the ethanol is steamed under reduced pressure. The residue is extracted with ether to remove the unsaponifiable components. The alkaline liquid is then acidified and extracted with ether. The obtained ether extract is then treated with cold dilute alkaline solution to remove the saponified acidic components, while coumarins are retained in the ether solution, precipitated, and crystallized after concentration of ether (Yu et al., 2022a). If the solution isn't directly precipitated and crystallized, vacuum distillation is performed, so they are divided into several fractions to be crystallized.

4. Physical and Chemical Properties of Coumarins

4.1. Dissolution and Fluorescence Properties

Most of free coumarins are crystalline colorless or light-yellow solid with a certain melting point, accompanied by aromatic odor. The smaller molecular weight of the volatile coumarins can be evaporated. Free coumarins are soluble in methanol, ethanol, chloroform, and boiling water, but not in cold water. Moreover, some coumarins show fluorescent properties and can display white or yellow crystals in the visible light (Lu et al., 2022). The fluorescence of hydroxycoumarin in alkaline environment is enhanced or even discolored. However, the fluorescence properties of hydroxyl groups are weakened or even disappeared after etherification. Notably, the fluorescence of furanocoumarins and pyranocoumarins are weak, while the fluorescence properties of coumarins with amino groups at the C-7 position are stronger than 7-hydroxycoumarin.

The structure of coumarins is a double ring, and their multiple double bonds increase the rigidity and conjugation of molecules, which makes coumarins possess strong fluorescence emission and good photostability. Other groups can also be introduced to coumarins at C-3, 4, 6 or 7 positions (Sun et al., 2022). Partial substitution of coumarins leads to blue shift or red shift of fluorescence. Therefore, the optical properties of coumarins can be adjusted in this way to obtain the desired target molecules. For example, the electron donor group at C-7 position is usually more conducive to enhancing the fluorescence emission, and the strength of the electron donor group can also influence the fluorescence quantum yield of coumarins by affecting the push-pull effect of electron in the rigid coplanar geometry. Moreover, coumarins have a transition between the excited state and the non-fluorescent distorted state in the plane, where this inhibition of the distortion process greatly affects the fluorescence quantum yield.

Coumarins and their derivatives exhibit high fluorescence quantum yield, great photostability and chemical stability, adjustable emission wavelength, and sensitivity to polarity of microenvironment (Kowalska et al., 2022). In addition, the advantages of available synthesis and modification on coumarins are also more targeted to obtain desired molecules. By connecting active groups of coumarins with other functional groups, the fluorescence signal can be changed by enhancing or inhibiting original electron transfer processes, isomerization, or other effects in the molecules. At present, many fluorescent probes based on modified and extended coumarin dyes have been designed and synthesized. The applications of these new coumarin materials are greatly extended to fields such as two-photon excitable fluorescent materials, organic light-emitting diode materials and solar cells (Luo et al., 2022).

4.2. Properties of Lactone

Coumarins show α and β-unsaturated lactone, so their structures have the generality of lactone compounds. In alkaline solution, the lactone ring can be slowly hydrolyzed and opened to form cis-o-hydroxycinnamate (Yang et al., 2022a). Cis-o-hydroxycinnamate is extremely unstable, and its aqueous solution can be closed-loop after acidification. It is reduced to the original lactone structure. Accordingly, coumarins can be extracted and separated by this way. Due to the existence of lactone ring, coumarins can perform a series of reactions through the cracking of lactone ring under alkaline conditions (Feng et al., 2022). The decarboxylation reaction and the formation reaction of new lactone ring on coumarins are shown in Figure 6 and Figure 7, respectively.

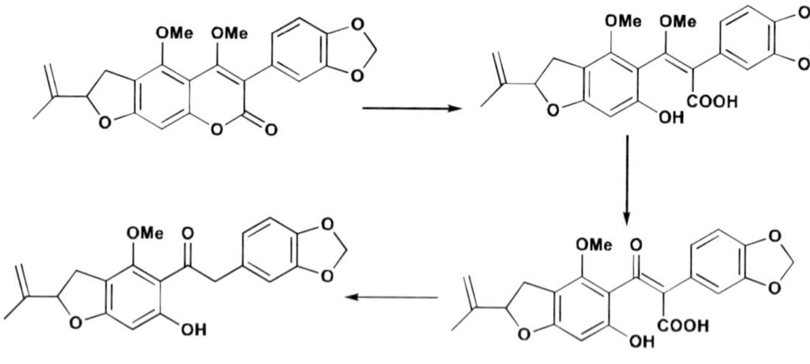

Figure 6. The decarboxylation reaction of coumarins.

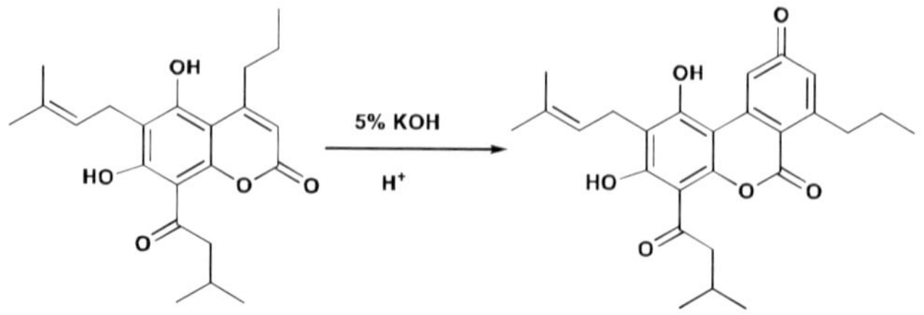

Figure 7. The formation reaction of new lactone ring of coumarins.

4.3. Oxidation Reaction

(1) Reaction with potassium permanganate
The oxidation capacity of potassium permanganate is relatively strong. When coumarin compounds are oxidized by it, the lactone ring is broken and salicylic acid derivatives are obtained (Figure 8).

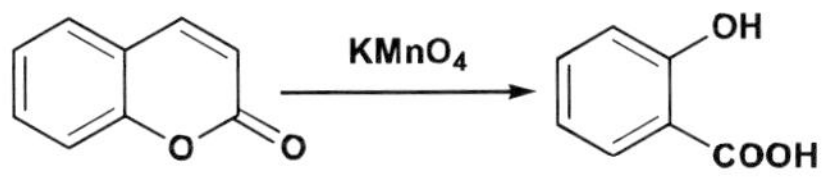

Figure 8. Reaction of coumarin with potassium permanganate.

(2) Reaction with chromic acid
Chromic acid as an oxidant shows mild properties and only oxidizes side chains or benzene rings. For example, osthole is oxidized only on the double-bonded side chain, obtaining a carboxylic acid (Figure 9).

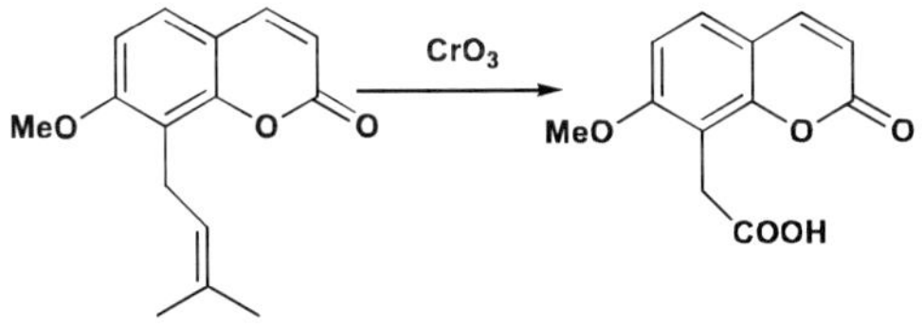

Figure 9. Reaction of osthol with chromic acid.

(3) Reaction with ozone
Ozone firstly reacts with the double bond on the side chain of coumarins, then reacts with the bond on furan or pyranone ring, and finally reacts with the double bond on α-pyranone ring under severe conditions (Figure 10).

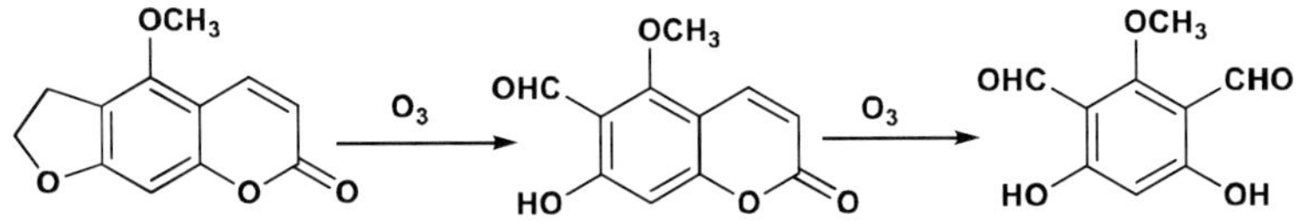

Figure 10. Reaction of furanocoumarins with ozone.

(4) Light reaction
Coumarin compounds have electron deficient double bonds at C-3 and C-4 positions, so they can form hydroxyl complexes with electron rich olefins

through photoinduced electron transfer, and then photoaddition reaction is carried out.

4.4. Substitution Reactions

(1) Reaction with acid
Under acidic catalysis, the double bond of the side chain on coumarin molecules can be hydrated to introduce hydroxyls (Figure 11).

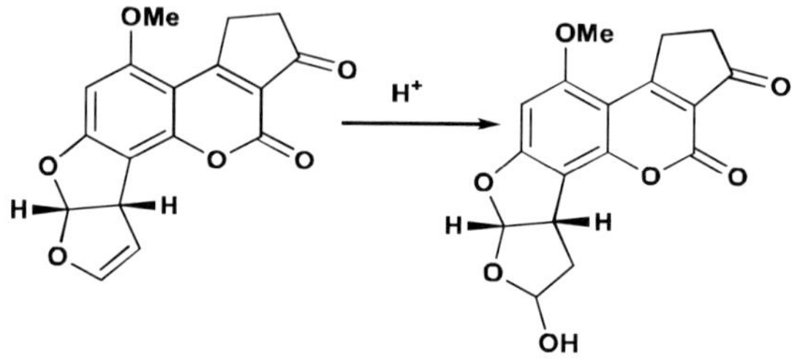

Figure 11. Reaction of coumarins with acids.

(2) Addition reaction of the double bond
The addition of coumarins can be divided into two types. One double bond on the mother nucleus and the other is on the side chain. Under controlled conditions, the double bond on the side chain is generally hydrogenated firstly, then the furan ring or pyran ring, and finally the hydrogenation on the coumarin ring is occurred (Figure 12).

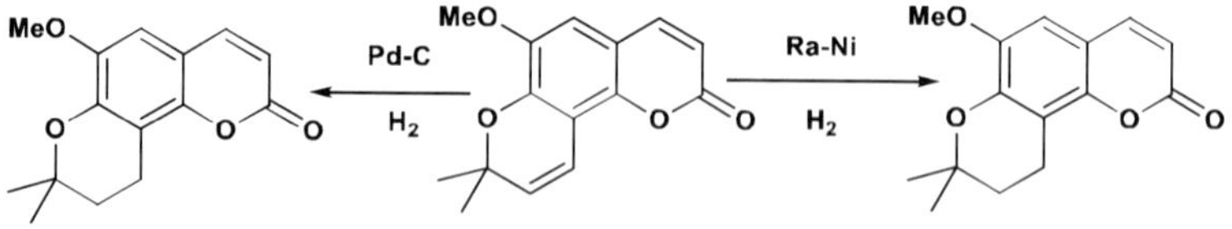

Figure 12. Addition reaction of coumarins.

5. Synthesis of Coumarins

Although coumarins are widely present in many plant sources. They can be prepared by solvent extraction, steam distillation, and acid precipitation, leaching, crystallization and extraction, but there are still many defects, such as low yield, long preparation time, cumbersome operation, and scarce raw materials. These factors limit the development and application of coumarin. Considering that the structures of coumarins are easy to be chemically modified, these problems can be solved by chemical synthesis. Nowadays, remarkable achievements have been made in the synthesis of coumarins and their derivatives.

There are many methods to synthesize coumarins, and the common traditional methods involve Perkin synthesis, Pechmann synthesis, Wittig synthesis and Knoevenagel synthesis (Heravi et al., 2014; Valizadeh and Vaghefi, 2009). Most of these methods use salicylaldehyde and resorcinol as starting materials, such as Perkin and Knoevenagel, which have successfully completed the total synthesis of coumarins by replacing salicylaldehyde. Pechmann et al. also successfully synthesizes coumarins by using resorcinol instead of salicylaldehyde as the starting material (Rodriguez-Dominguez and Kirsch, 2006). Additionally, Wittig synthesis and the reaction of proparynic acid derivatives are also considered as effective methods for the synthesis of coumarins. Traditional synthesis ways usually show shortcomings. Therefore, simple, efficient, and green new synthesis methods emerge at the historic moment. Such as Mukaiyama esterification, ionic liquid chlorination trimethylhexadecylamine catalysis, ultrasonic, microwave, etc.

5.1. Total Synthesis of Coumarins

5.1.1. Perkin Synthesis Method
Perkin synthesis is described in Figure 13 (Aslam et al., 2010). The coumarin parent nucleus of pyranone ring is synthesized from o-hydroxybenzaldehyde) and acetic anhydride at 180 °C under the catalysis of potassium acetate, but the reaction time is long, the reaction temperature is high and the yield is very low. Through continuous optimization to change the reaction conditions and material ratio, the reaction yield has been significantly improved. For example, KF is used as a catalyst, and an appropriate amount of phase transfer catalyst is added to the reaction, which greatly reduce the number of by-products. Moreover, coumarins are also synthesized by microwave in another experiment (Marriott et al., 2012). The experimental results show that the yield of coumarin can reach 92% when the molar ratio of salicylaldehyde: acetic anhydride: potassium carbonate is 1:3:0.2, the microwave power is 495 W and the reaction time is 5 min. The method has the advantages of high yield, short

reaction time and simple operation It is the first way for the synthesis of coumarin parent nucleus, and conventional for industrial production of coumarins, which has been used till now.

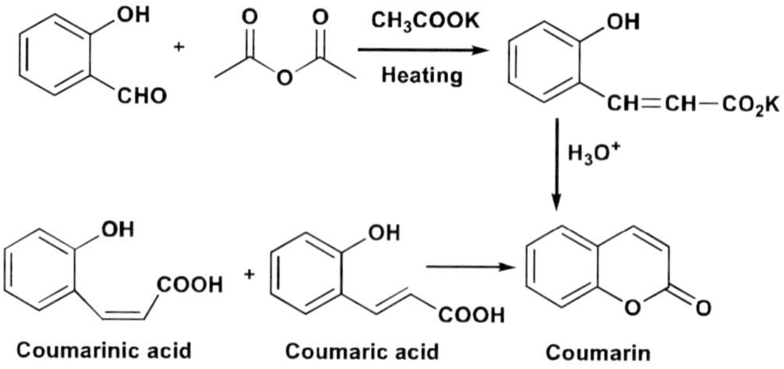

Figure 13. The total synthesis of coumarins by Perkin.

5.1.2. Pechmann Synthesis

Pechmann condensation is the most used way for the synthesis of simple coumarins (Potdar et al., 2001). In 1884, Pechmann, a German scientist, was firstly to synthesize coumarins in a closed loop through the reaction of resorcinol and malic acid under the catalysis of dried concentrated sulfuric acid at 120 °C (Chaudhari, 1983). By the later improvement of the method, coumarins are obtained by condensation reaction of substituted phenol and β-ketoate ester with Lewis acid catalysis, which is the best way for the preparation of substituted coumarins on benzene ring (Figure 14).

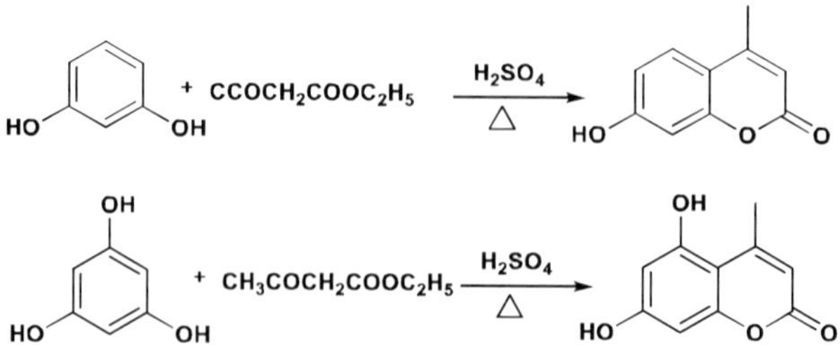

Figure 14. Pechmann synthesis of coumarins.

The advantages of Pechmann condensation are easy to obtain raw materials and mild reaction conditions. The disadvantages are that it requires a large amount of Lewis acid or proton acid, with many by-products and low yield. Even more a lot of waste water is produced after treatment, which has adverse effects on the environment. Moreover, many phenols do not have this reaction, and furanocoumarins cannot be prepared by this way, because furan ring is too sensitive to acid. In view of this, many researchers have done a lot of improvement work. Various new catalysts and technologies have emerged in endlessly and achieved good results. The newly developed catalysts include $InCl_3$, $Bi(NO_3)_3$, $TiCl4$, micro zinc, molecular sieve, etc. (Karatas et al., 2021), which greatly improves the yield. With new technologies rising, Pechmann reaction in neutral ionic liquid [Bmim]PF6 or [Bmim]PF4 without catalyst is carried out, and the yield is higher than 90% (Gopalakrishnan et al., 2009). Among them, microwave technology has greatly improved Pechmann reaction in terms of reaction time. For example, five kinds of 4-substituted coumarins catalyzed by concentrated H_2SO_4 are synthesized in a household microwave oven at 200 W. The reaction time is only 2–10 min, and the yield is increased to 68–82% (Xu et al., 2018). Pechmann condensation is widely used. It can not only use substituted phenol as raw material to synthesize coumarins, but also directly introduce various substituents on the pyranone ring, which greatly expands the types of coumarins.

5.1.3. Wittig Synthesis

Wittig reaction is often used to synthesize coumarins without substituents at C-3 and C-4 positions (Mustafa et al., 2020). Phosphorus ylide is synthesized from ethyl chloroacetate and triphenylphosphine under alkaline conditions. Then, Wittig reagent is generated after phosphorus ylide removing the hydrochloric acid. Immediately, substituted o-hydroxybenzaldehyde reacts with Wittig reagent to remove a molecule of ethanol and then cyclizes to form coumarin parent nucleus (Figure 15).

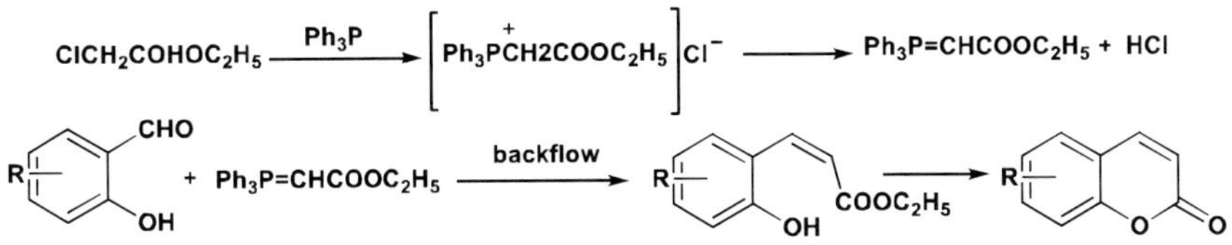

Figure 15. Witting synthesis of coumarins.

An efficient and chemically selective one-pot method for the synthesis of two coumarins based cross-coupling adducts is demonstrated, namely 2,3-furanocoumarin and 3-benzofuranocoumarin (Li et al., 2022a). The key to the method is the chemical selectivity of the functionalized amphoteric ion and acetylation reagents as well as selective Wittig reaction in molecules. The intramolecular Wittig reaction and the addition sequence of acetylation reagents determine the yield of two products (Figure 16).

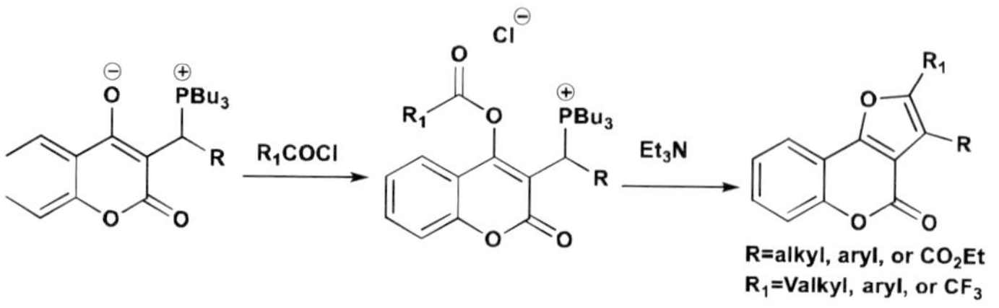

Figure 16. Witting synthesis of furancoumarins.

5.1.4. Knoevenagel Synthesis

Knoevenagel optimizes the Perkin reaction. Under the catalysis of weak base, 3-substituted coumarins are synthesized by dehydration condensation of o-hydroxybenzaldehyde or ketone and acetic acid derivatives with α-H atom (Xie et al., 2012). Knoevenagel reaction is suitable for the participation of electron-substituted salicylic aldehyde. 7-hydroxy 3-coumarin ethyl formate is synthesized by using 7-hydroxysalicylic acid and diethyl malonate in the presence of piperidine, together with ethanol as solvent and reflux for 2 h (Coulibaly et al., 2022). The yield is up to 97%. In another experiment, a small amount of water is added to 1,3-dimethylimidazolium methyl sulfate as the reaction solvent and catalyst, and Knoevenagel condensation reaction occurs at room temperature. It takes 2–7 min to complete, and the yield is up to 92%–99% (Figure 17) (Verdia et al., 2017).

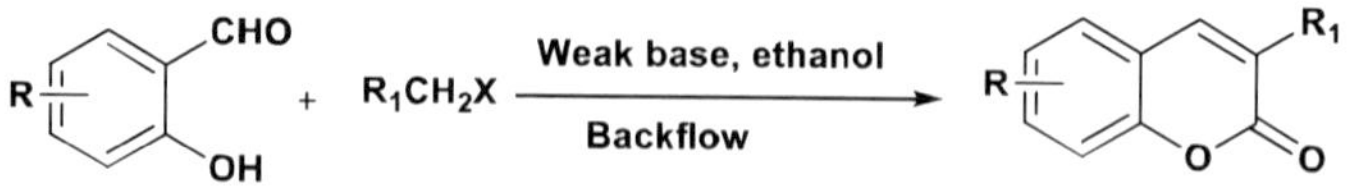

Figure 17. Knoevenagal synthesis of coumarins.

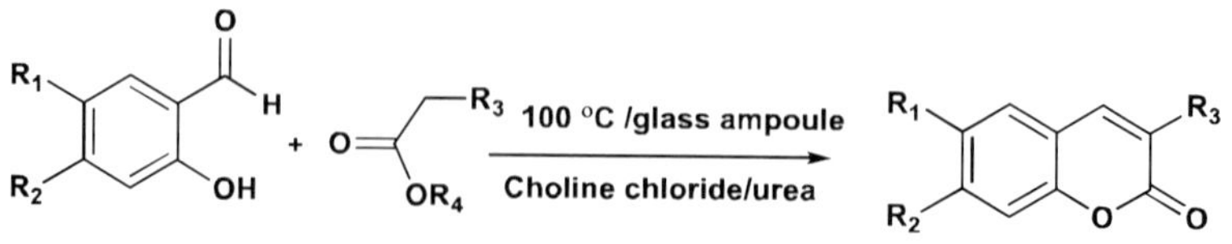

Figure 18. Optimized Knoevenagel synthesis of coumarins.

The reaction conditions by Knoevenagel method are further improved. The coumarin derivative is obtained by reaction of Meldrum's acid with choline-urea chloride ionic liquid instead of salicylic aldehyde under solid state condition with 95% yield (Figure 18) (Xi and Liu, 2015).

5.1.5. Remier-Tiemalni Synthesis

As shown in Figure 19, the synthesis of coumarin by Remier-Tiemalni method is completed in two steps (Barriga-Gonzalez and Munoz-Espinoza, 2022). Firstly, salicylic aldehyde is synthesized from benzene and chloroform under the action of sodium hydroxide. The second step is to synthesize coumarin from acetic anhydride and salicylaldehyde with sodium acetate.

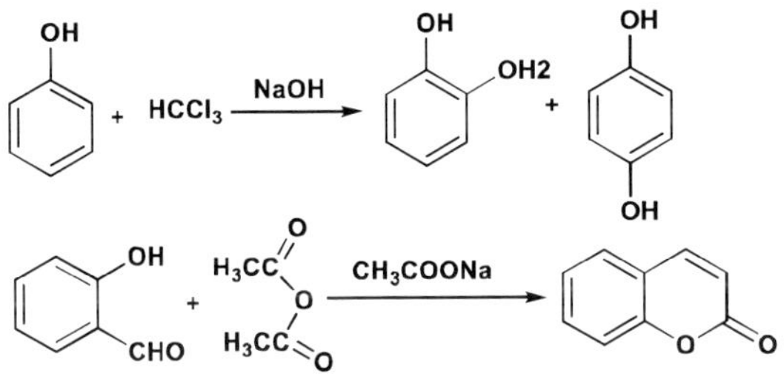

Figure 19. Remier-Tiemalni synthesis of coumarins.

5.1.6. Sodium Chloroacetate Synthesis

The synthesis of coumarin skeleton by sodium chloroacetate method is completed in three steps. Firstly, chloroacetic acid is substituted with sodium hydroxide to produce sodium chloroacetate. Secondly, the carboxyl coumarin is then synthesized with salicylic aldehyde under the catalysis of sodium hydroxide and sodium cyanide. Eventually, free coumarins are prepared by decarboxylation and finally by vacuum distillation. The yield of coumarin synthesized by sodium chloroacetate can reach 80% (Yang et al., 2018). Compared with other preparation methods, coumarin synthesized by this way reveal the advantages of high yield and less by-products. The reaction synthesis route is shown in Figure 20.

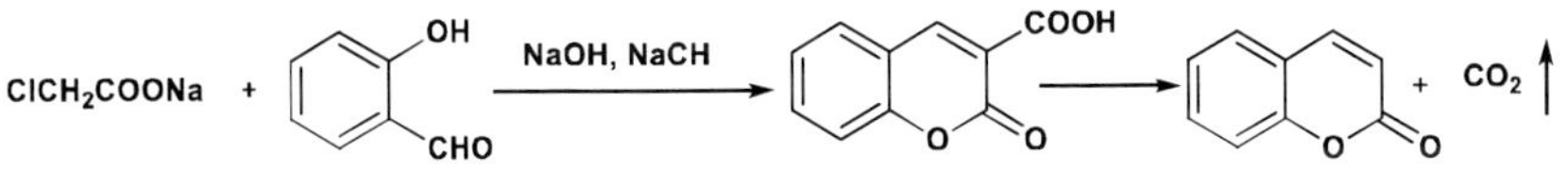

Figure 20. Sodium chloroacetate synthesis of coumarins.

5.1.7. Vilsmeier-Haack Synthesis

Vilsmeier-Haack method is also convenient to synthesize coumarins. With N,N-dimethylformamide as a solvent, salicylic aldehyde and N,N-dimethyl-substituted acyl chloride under the action of perchloric acid prepare the coumarin skeleton via nucleophilic addition and cyclization. The reaction synthesis route is shown in Figure 21 (Pushpalatha et al., 2016).

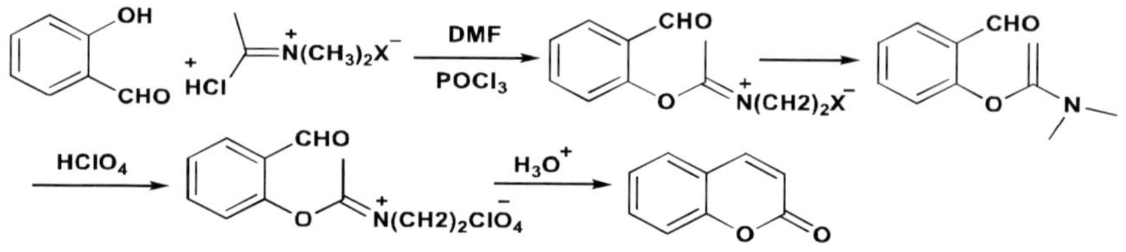

Figure 21. Vilsmeier-Haack synthesis of coumarins.

5.2. Synthesis of Coumarin Derivatives

Coumarins as organic intermediates don't possess abundant bioactivities and excellent fluorescence properties. However, according to their flexible structural design and easy modification, many coumarin derivatives with different structures, properties and bioactivities can be synthesized by introducing modified groups at different substitution sites on the coumarin skeleton.

5.2.1. Modification on Coumarins at C-3 Position

The C-3 position of coumarins has always been a hot spot for structural modification. Many studies have found that the amide bond at C-3 position on the B ring of coumarin can enhance their antitumor activity, as described in Figure 22 (Williams and Gieling, 2019). Here, the coumarin derivative 32 and the quinoline-3-amide derivative 33 are designed and synthesized. The results show these compounds exhibit certain proliferation inhibition activities against prostate cancer cells, colon cancer cells, breast cancer cells, etc. Among them, compound 34 shows high selectivity to prostate cancer cells with GI_{50} less than 10 μM. Meanwhile, the compound 32 with coumarin as the nucleus reveals better anticancer activity than the compound 33 with quinoline, suggesting that coumarin as the nucleus plays an important role in antitumor activity.

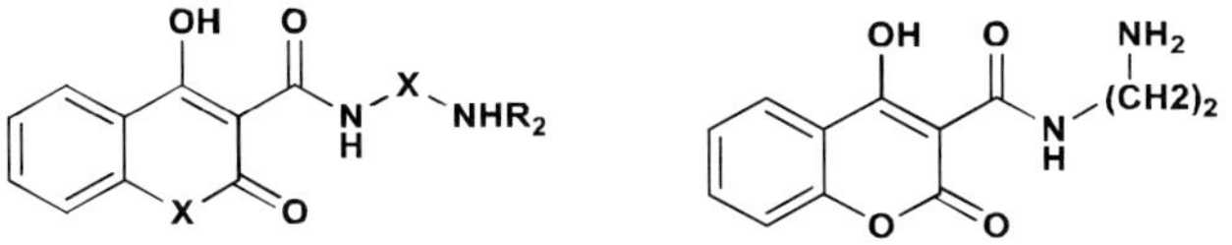

Chemical Constituents and Synthetic Methods ... 43

Figure 22. Structures of Coumarin Derivatives 32, 33 and 34.

Hydrazizone group as a ligand can enhance the anti-tumor bioactivity of coumarins, and the introduction of furan, thiophene, pyrrole and other skeletons can also improve their bioactivities. Accordingly, coumarin hydrazide derivatives 35–37 are successfully synthesized (Fuentes-Aguilar et al., 2022). Among them, compounds 35 and 36 have better activity against drug-resistant pancreatic cancer cell than doxorubicin. The compound 36 also shows good inhibitory activity on selected tumor cell lines.

5.2.2. Modification on Coumarins at C-4 Position

Most coumarin derivatives substituted at the C-4 position have good pharmaceutical activities. For example, 4-hydroxycoumarin is an intermediate that can be used as perfume and dye, and be applied in the synthesis of nitrobenzyl coumarin, dicoumarin, warfarin and anticoagulant drugs. 4-hydroxycoumarin also has important application value in anti-cancer, anti-fungal and treatment of cardiovascular diseases. It can be synthesized by the methyl salicylate way. Methyl salicylate and acetic anhydride are used as raw materials and concentrated sulfuric acid is as catalyst to prepare 4-hydroxycoumarin via esterification and cycination. The reaction synthesis route is shown in Figure 23 (Chatterjee et al., 2019).

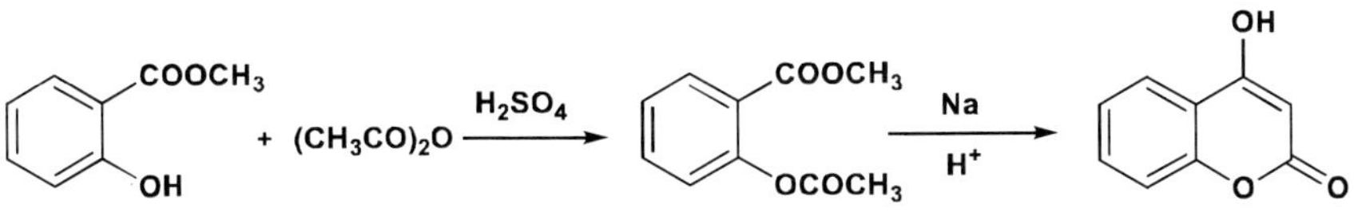

Figure 23. Synthesis of 4-hydroxycoumarin by methyl salicylate.

Cyclin phosphorylase is an important phosphatase regulating cell cycle, which inhibits cell proliferation by regulating the dephosphorylation of amino acid residues on cyclin-dependent kinases/cyclins complex. Compounds **40** and **41**, whose B ring on coumarin parent nucleus, is replaced by the 4-benzoyl ethylene group, showing the best inhibitory activity against CDC25 with inhibition rates of 94.2 ± 7.3% and 94.3 ± 7.9%, respectively. However, compounds **42** and **43**, whose carbonyl group is directly connected to the parent nucleus of coumarin, have no inhibitory activity against CDC25 (Bana et al., 2015). These compounds are illustrated in Figure 24.

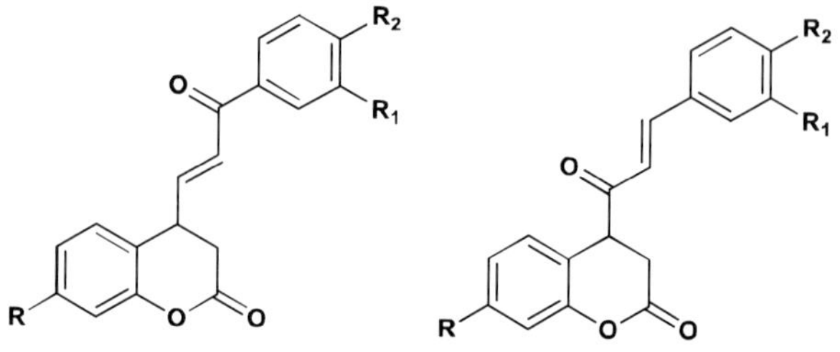

Figure 24. Structures of coumarin derivatives 40, 41, 42, and 43.

5.2.3. Modification on Coumarins at C-6 Position

The substituted coumarin derivatives at C-6 position are mainly used as functional intermediates in organic synthesis and medicine as well as spices. Among them, 6-methyl coumarin is the representative compound. It is prepared by alienation, decarboxylation and cyclization reaction using *p*-methylphenol and *trans*-butenic acid as raw materials and concentrated sulfuric acid as catalyst (Wang et al., 2017). Figure 25 shows the synthesis route.

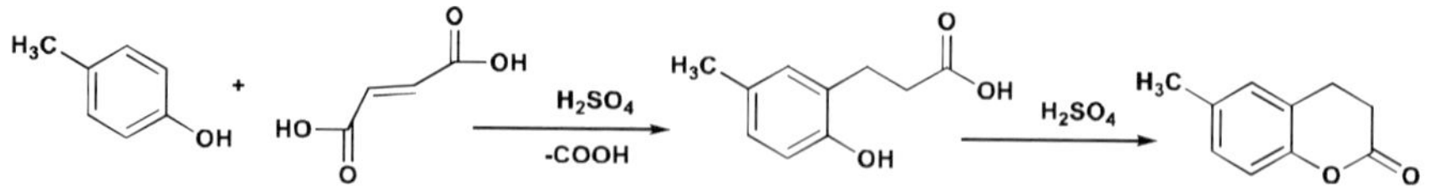

Figure 25. Synthesis of 6-methyl coumarin.

Resveratrol is abundant in plants and has many bioactivities such as hypoglycemic and anti-tumor activities. It has a good inhibitory effect on the formation and proliferation of tumor cells, but its low bioavailability and rapid plasma clearance make it difficult to be used as a drug. Considering these problems, 18 compounds of resveratrol and coumarin are synthesized. Among them, compounds 46-49 (Figure 26) have good inhibitory activities on MCF-7, HCT-28 and K562 (IC_{50} value range 3.78–19.16 μM) (Shen et al., 2013).

Compound 46 shows the best inhibitory activity against HCT-28 with IC_{50} 3.78 μM. Compound 47 has the best inhibitory activity against K562 with IC_{50} 3.79 μM. Compound 49 shows the highest inhibitory activity against MCF-7 with IC_{50} 4.23 μM. Structure-activity studies have found that only compounds with A and C rings linked by double bonds or other groups in trans conformation can bind to the receptor and exhibit proliferation-inhibiting activity.

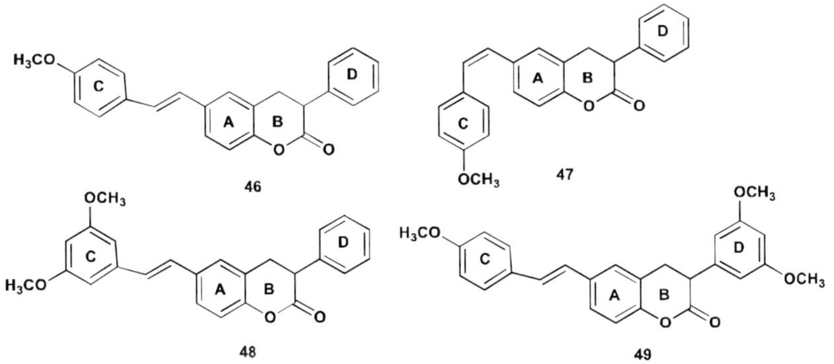

Figure 26. Structures of coumarin derivatives 46–49.

5.2.4. Modification on Coumarins at C-7 Position

Coumarin derivatives substituted at the C-7 position are mainly used as pharmaceutical intermediates. 7-hydroxyl-4-methyl coumarin is an important derivative, which is prepared by cyclization reaction using resorcinol and ethyl acetoacetate as raw materials and concentrated sulfuric acid as catalyst, as shown in Figure 27 (Lamani et al., 2009).

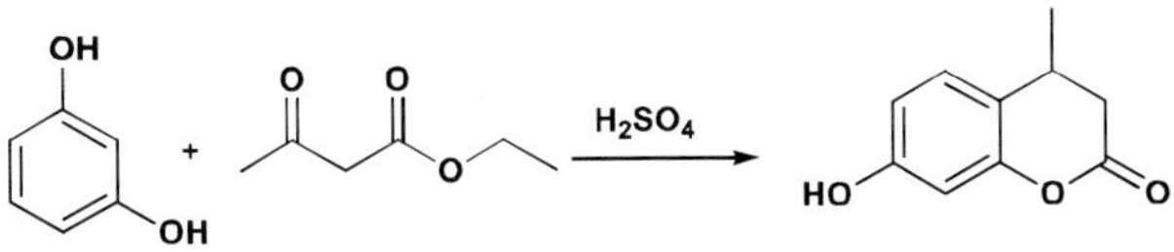

Figure 27. Synthesis of 7-hydroxyl-4-methyl coumarin.

In other experiments, a series of coumarin derivatives were designed and synthesized by introducing different groups at C-7 position of coumarin with O and S atoms as connecting groups (Xia et al., 2015). Among them, compound 53 shows certain anti-proliferation activity against A549 and

DU145, and inhibited the proliferation of A549 by inducing apoptosis and arresting the cell cycle in G2/M phase. Moreover, compound 54 selectively inhibited MDR KB VIN in tumor cells and show high bioactivity. The substituent group at C-7 position selectively combines with the action target (Kosenko et al., 2020). As for these compounds, it is particularly important to introduce the substituent group with appropriate size at the 7-position.

5.2.5. Modification on Coumarins at C-8 Position

According to the structure-activity relationship of drugs, the introduction of pyrazole groups can enhance the bioactivity of drugs, which is of great significance in drug design. For example, compound 57 containing pyrazole groups shows good selectivity and proliferation inhibition activity against tumor cells (Debbabi et al., 2016). Several coumarins substituted by pyrazole group at C-8 position are synthesized, and all of them reveal good proliferation inhibition activity against MCF-7 and HCT-116. Especially, compound 60 shows inhibitory effects on 60 tumor cell lines with IC_{50} values ranging from 0.9 μM to 4.8 μM. Compounds 58, 59 and 60 displays good inhibitory activity against HCT-116 with IC_{50} values of 0.02 μM, 0.02 μM and 0.01 μM, respectively (Detsi et al., 2017). These coumarin derivatives with pyrazole groups are shown in Figure 28.

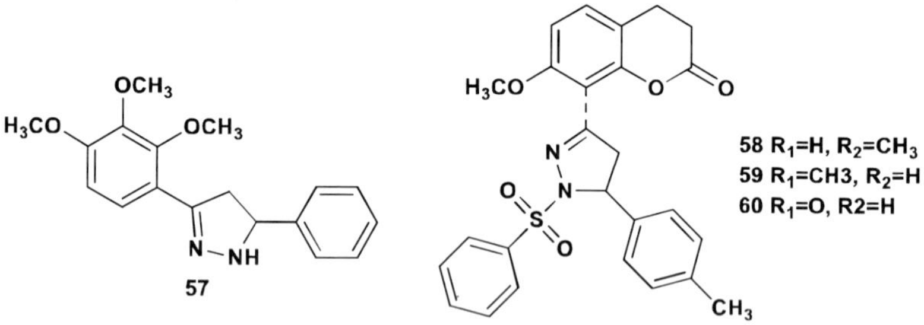

Figure 28. Structures of coumarin derivatives with pyrazole groups (compounds 57–60).

Gallic acid, a natural product, can prevent cell cycle and inhibit the proliferation of tumor cells. 8 compounds are prepared by introducing gallic acid at the C-7 position and substituting different groups at the C-8 position through the benzene ring of coumarins (Li and Seeram, 2010). It is found that the different substituents at the C-8 position have a great influence on the activity of the compounds. Among them, compound 61 (Figure 29) shows good selectivity and inhibitory activity against DU145, and the cytotoxicity is particularly low. Excitedly, the side effects on normal cells are minimal.

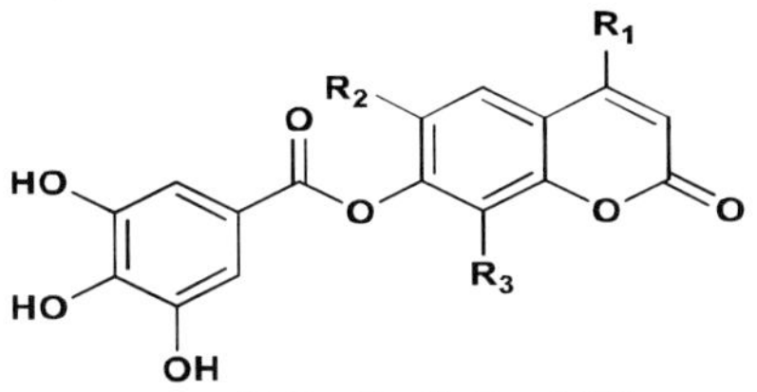

Figure 29. Structures of compound 61.

5.3. Biosynthesis of Coumarins

The biosynthesis of coumarins is a branch of the pathway of phenylpropanoid biosynthesis. The biosynthesis of phenylalanine begins with the formation of coumaryl-coA through a multi-step reaction (Sui et al., 2019). Phenylalanine diaminylase is a key enzyme in this process and a rate-limiting enzyme in the biosynthesis of most phenylalanine compounds. The biosynthesis of phenylpropanoid compounds in plants is mediated by the expression of phenylpropane-related genes, which are regulated by various transcription factors and signaling molecules. Finally, the enzyme protein with catalytic activity related to phenylpropyl metabolism is synthesized. These enzyme proteins achieve the synthesis and modification of phenylpropanoid compounds through their unique spatial structure and activity (Zhao et al., 2021). However, studies on the biosynthesis mechanism of coumarins in plants is still a worthy of further investigation. Caffeioyl coenzyme A methyltransferase is related to the synthesis of lignin. When the coding gene is knocked out, the synthesis of hyoscyamine will be significantly reduced. This suggests that ferulic acid coenzyme A may be the precursor of coumarin synthesis.

Some scholars believe that the synthesis of coumarins originates from cinnamic acid, and the enzymes related to cytochrome P450 exert an important effect during the synthesis. The o-hydroxylation of cinnamic acid which occurs in chloroplasts is a key step. The hydroxylation process is dependent on cytochrome P450 (Niwa et al., 2022). As experiments are elaborated as follows. The biosynthesis of scopoletin in Arabidopsis is closely related to cytochrome P98A3 enzyme, which mainly catalyzes the glycosylation reaction in P-coumaric acid. In another exploration, metabolic regulation related to hydroxycoumarin biosynthesis is investigated by isotope labeling. The results indicate that the generation of scopoletin and its glycosides are achieved through the sequence of P-coumaric acid, caffeic acid and ferulic acid (Langat et al., 2014). Yet, the synthesis of aesculin and its glycosides are realized by p-coumaric acid to 2', 4'-dihydroxy carnophytic acid and then to 7-hydroxy coumarin. The synthesis of these coumarins mainly involves hydroxylation. A new biosynthetic pathway of 4-hydroxycoumarin is discovered by gene cloning. With the help of biphenyl synthase, the isoenzymes encoded by them can catalyze the decarboxylation reaction with malonyl-coA using o-hydroxybenzoyl-coA as the substrate, and finally 4-hydroxycoumarin is obtained (Liu et al., 2010). At present, there are only a few reports on the biosynthetic pathway of plant coumarins, including simple coumarins and furan coumarins. Unfortunately, studies on the biosynthesis mechanism of pyranocoumarins with complex structure and common coumarins containing multiple methoxy groups have not been reported.

5.3.1. Biosynthesis of Simple Coumarins

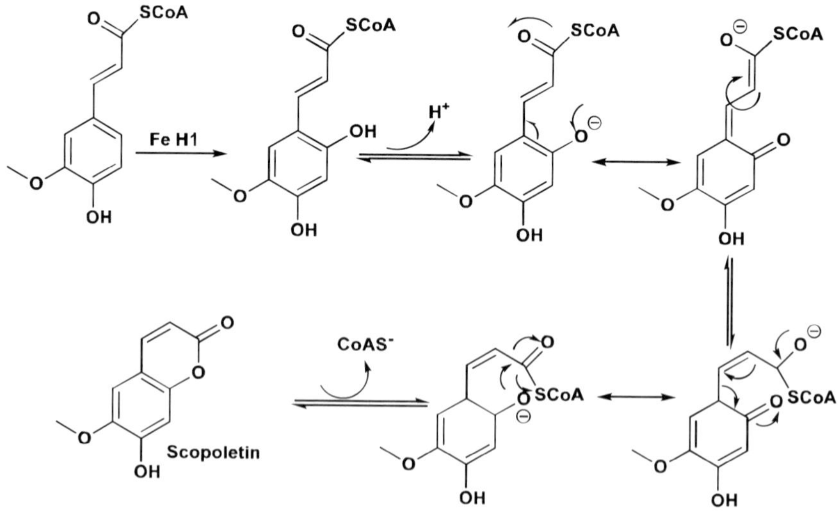

Figure 29. Biosynthetic pathway of scopolamine in Arabidopsis in *vivo*.

Due to many studies on the biosynthesis of furacoumarins, most scholars believe that cytochrome P450 is also essential in the biosynthesis pathway from benzo esters to coumarins. Japanese scholars have studied the biosynthetic pathway of scopolamine in Arabidopsis. Experiments show that coenzyme A and esters are required as substrates to hydroxylate under the catalysis of soluble Fe(II)-2-ketoglutarate dependent enzymes to form coumarin compounds (Chen and Walsh, 2001). The specific pathway is shown in Figure 29.

5.3.2. Biosynthesis of Furanocoumarins

Furanocoumarins are one of the most studied compounds in the biosynthesis of coumarins. Studies have shown that the synthesis of furanocoumarins is related to cytochrome P450, which is formed by the hydroxylation of the isopentenyl group at the C-66 or C-8 position of umbelolactone. For example, psoralen synthase has been cloned and identified as a key enzyme in psoralen formation. The specific synthetic steps of furancoumarins are shown in Figure 30 (Larbat et al., 2009).

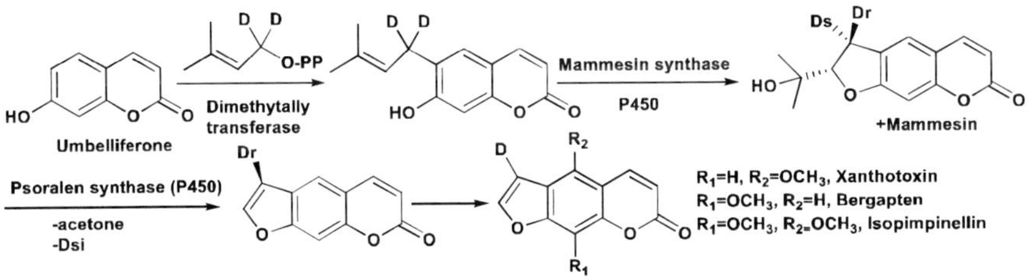

Figure 30. The transformation process of furacoumarins in plants.

6. Bioactivities of Coumarins

Natural coumarins are widely distributed in animals, plants, and microorganisms. Due to their small molecular weight, π-π conjugation system and rigid viscous ring, they can act on various active sites in organisms through non-covalent interaction (Supuran, 2020). Generally, coumarins are obtained from natural resources, biological and chemical synthesis. Coumarins and their derivatives have high medical value. The antitumor mechanisms of some coumarins have been evaluated at the molecular level. Warfarin sodium is the first coumarin drug used for cancer treatment, which has high anticancer activity (Huang et al., 2017). In addition, 7-hydroxycoumarin influences kidney cancer cells, but it has a certain selectivity. Pyranocoumarins can inhibit HIV-1 to a certain extent (Huang and Liu, 2018). Briefly, pharmacological tests show that coumarins reveal many biological activities, such as antiviral, anti-cancer, anti-microbial, anti-inflammatory, antioxidant, etc. (Rostom et al., 2022).

6.1. Anticancer Activity

Cancer is one of the leading causes of death for many people, and researchers are trying to find new cancer drugs. Herein, 15 coumarins are screened out to act on MCF-7, SW48, HT-29, and A549 cancer cell lines, as well as two normal cell lines 3T3 and HUVEC. It is found that some groups on the benzene ring have toxic effects on cancer cells *in vitro* by inducing apoptosis, and 1,5-dihydropyran [2,3-c] benzopyran derivatives can be the leading compounds in the synthesis of anticancer drugs (Bansal et al., 2013). These coumarins have certain anti-tumor effects and little toxicity to normal cells. Meanwhile, recent experimental results show that a variety of coumarin derivatives show similar anti-tumor activity to 5-FU on tumor cells (Kuchlyan et al., 2014; Wang et al., 2015).

In this experiment, several coumarin derivatives with benzimidazole are synthesized and their anticancer activities are evaluated *in vitro* (Karatas et al., 2019). The results show that these derivatives reveal excellent inhibitory activity on human gastric cancer cells. Even after the benzimidazole structure is methylated, their anticancer activities are significantly enhanced. When substituted by methyl or methoxy, coumarin derivatives exhibit strong inhibitory effect on cervical cancer cells. As substituted by methyl or bromine, coumarin derivatives exhibit superior inhibitory effect on human colon cancer cells. Except for imidazole substituents, coumarin derivatives of thiazolyl Schiff base with anticancer activity are prepared (Kumar et al., 2021). Coumarins with 2,6-dichlorophenyl structure show good inhibitory activity on Hela tumor cells. Interestingly, coumarins with 2-chloro-5-nitrophenyl and 10-chloro-anthracene have good inhibitory effect not only on Hela tumor cells, but also on COS-7 tumor cells. Both of which are better than the positive control doxorubicin.

As found in other experiments, 4-anilinocoumarin derivatives substituted at the C-3 position have certain inhibitory effects on MCF-7, HepG2, HCT116 and PANC-1 cancer cells (Kaur et al., 2016; Luo et al., 2017). When the hydrogen on the amino group of the compounds is replaced by benzodioxin and 2-hydroxy-3-methylphenyl, their inhibitory activities are better than that of the positive control 5-FU. Coumarin isoxazoline compounds (Figure 31) have also been obtained, and these derivatives can well inhibit human melanoma cells (Lingaraju et al., 2018). Some of these compounds have significant antiproliferative effects on human melanoma.

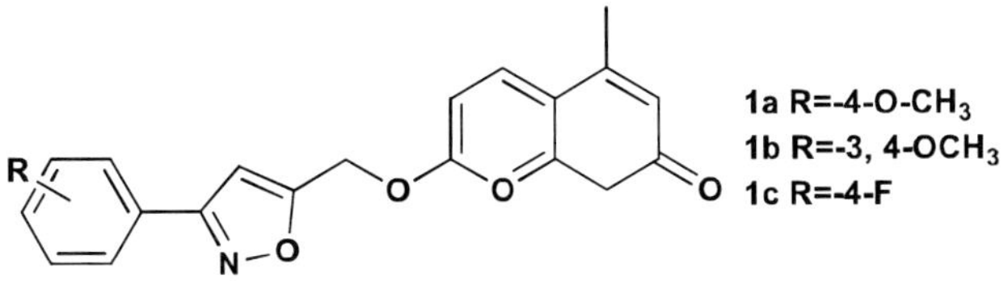

1a R=-4-O-CH$_3$
1b R=-3, 4-OCH$_3$
1c R=-4-F

Figure 31. Structure of coumarins with isoxazoline.

A variety of heterocyclic derivatives of coumarins have also been obtained. Results of cytotoxicity test *in virto* show that some compounds have good inhibitory activity against Ehrlich ascites cancer cells and Dalton lymphoma cells, which is close to the activity of 5-FU as the positive control (Ibrahim et al., 2022b). Nifurtimox is a drug used to treat high-risk neuroblastoma. It has a strong regulatory effect on the signal pathway of BCL2 family in mitochondria. At the same time, as a cyclin inhibitor, it can increase the expression level of cycle related protein kinase p27 and decrease the expression level of cyclin-D1 and CDK-6 (Singh et al., 2011). OVCAR-3 cell cycle is blocked in G2 phase and cell proliferation is inhibited. At present, it is in clinical phase II study as a drug for the treatment of childhood neuroblastoma.

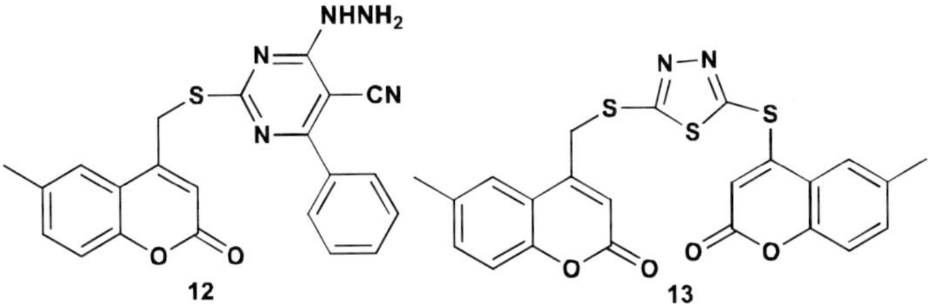

Figure 32. Structures of compounds 12 and 13.

In other experiments, 24 coumarin derivatives are obtained. Most of the compounds exhibit inhibitory effects on the proliferation of McF-7 and Hep G2, among which compounds 12 and 13 (Figure 32) have similar inhibitory activities to the positive control drug 5-FU (about 10 μM) (Li et al., 2021b).

Coumarin derivatives are generally very low toxicity. Compared with the stability of 5-FU and its disadvantages of not being used alone, this kind of compound has obvious advantages.

Novobiocin 8, a penicillin antibiotic, has strong effects on Gram-positive bacteria and Enterococcus faecalis (Llanos et al., 2022). It is reported that neomycin as a DNA helicase inhibitor binds to a nucleotide site at the C-terminal of heat shock protein in breast cancer cells, resulting in the proliferation of breast cancer cells (Ghosh et al., 2022). The neomycin analogue **9** is obtained by structural modification of neomycin. Its activity is significantly higher than that of neomycin, and its IC_{50} is 10 µM (Hanessian et al., 2011). Interestingly, natural products DHN1 and DHN2 have better anticancer activity than neomycin, with IC_{50} values of 7.5 µM and 0.5 µM, respectively (Zhao and Blagg, 2013). Neomycin analogues and natural products DHN1 and DHN2 are described as Figure 33.

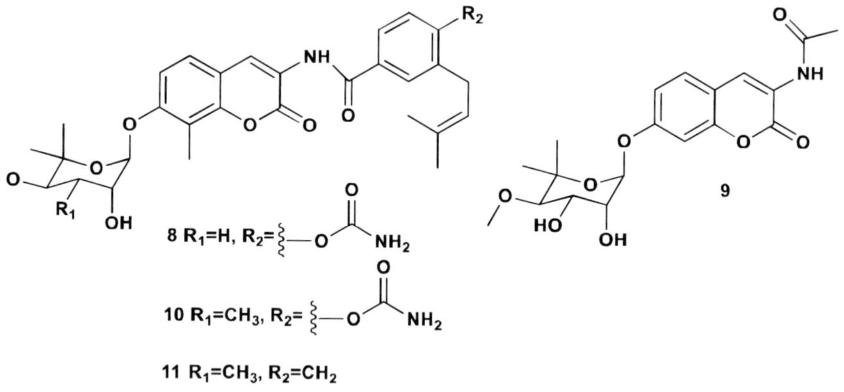

Figure 33. The structures of neomycin analogues and natural products DHN1 and DHN2.

In addition to the synthesis of coumarin derivatives, they can also be obtained from plants. Dracunculin (Figure 34) isolated from the Artemis plant is a simple coumarin containing a methyl, which can serve as a potential anti-tumor agent (Alamzeb et al., 2015). Xanthotoxin, bergapten and furanocoumarin curcumolide (Figure 34) are extracted from the root of Heracleum dissectum (Hosseinzadeh et al., 2019). Among them, furanocoumarin curcumolide is also found in the water extract of Peucedanum praeruptorum. Xanthotoxin is still purified from the ethanol extract of the stem of Salvadora indica (Iyer and Patil, 2014). These natural coumarin derivatives have good antitumor activity.

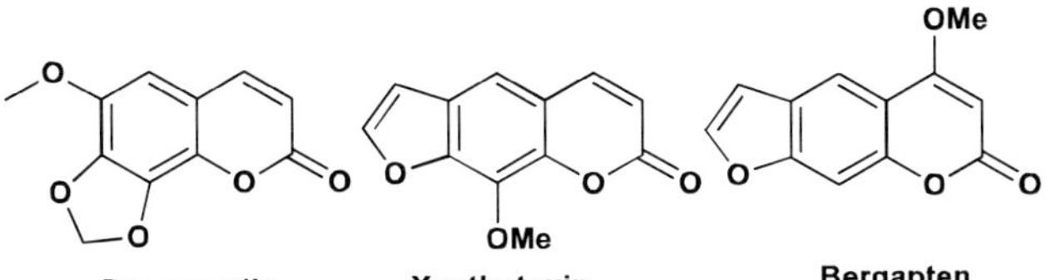

Figure 34. The structures of dracunculin, xanthotoxin, and bergapten.

The antitumor activity of coumarins is closely related to their chemical structure, which is affected by the difference of coumarin mother nucleus, the number and position of hydroxyl substitution, and the way of hydroxylation. The antitumor activities of functional groups on coumarins mainly include aryl at C-3, phenyl at C-4, alkyl at C-6, hydroxyl at C-7, etc. (Zhu and Jiang, 2018). It has been reported that the induction effect and conjugation effect of the coumarin ring can be increased by introducing larger groups at C-3 or C-4 position (Al-Majedy et al., 2017). Coumarins show obvious inhibitory activity on a variety of tumor cells when introducing aryl, furan, pyrrole, thiophene and other bulky groups at C-3 position. The antitumor activity of coumarins can also be improved by introducing o-hydroxymethoxy and o-hydroxy groups into coumarin rings while aryl groups are added at C-3 position. Furthermore, the direct introduction of benzimidazole and benzothiazole at C-3 position also has a positive effect on the anticancer activity of coumarins (Thakur et al., 2015). In the presence of amide bonds on coumarins, N-phenyl substitution strongly affects the activity of the compounds against tumor cells. The hydrazide group plays an important role in enhancing the antitumor drugs of coumarins. The presence of methylene at C-4 position of coumarin is the best bond for introducing large volume groups. By introducing aryl or aryl hetero groups into methylene, their antitumor activity can be significantly improved. If the introduced heteroaryl group is pyridine, the alkalinity of nitrogen on the pyridine ring is also important to improve the inhibitory activity of cancer cells (Akhtar et al., 2017). The antitumor activity of coumarins can be improved by introducing styryl with methoxy and methyl groups into the C-4 position of coumarins. Among many natural coumarins, simple 7-hydroxycoumarin derivatives have shown the required antiproliferative activity (Sandhu et al., 2014). If the sulfamate group or alkyl group is introduced at C-7 position, its antitumor activity can also be increased. Overall, the structure-activity relationship of coumarins is shown in Figure 35.

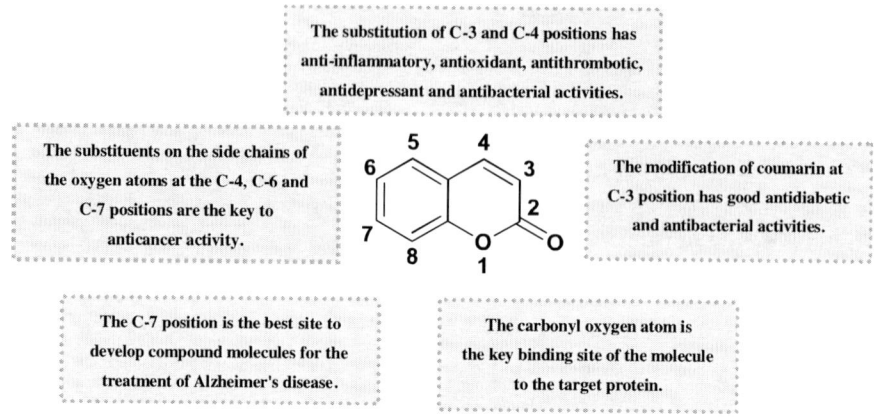

Figure 35. The structure-activity relationship of coumarins.

6.2. Antibacterial Activity

Natural coumarins usually exhibit good inhibition to a variety of bacteria. The methanol extract from the stems of *Daphane gnidiun L* has a good effect on Bacillus tardiness and Escherichia coli, but no inhibitory effect on fungi. 4. Coumarins are also obtained from the methanol extract, including daphnetin (1), daphnin (2), acetyl umbelolactone and dibertin (3). Additionally, 8 coumarins isolated and purified from *Tordylium apulum*, involving cnidiadin (4), imperatorin (5), isopimpinellin (6), xanthotoxin (7), etc. (Chen et al., 2007). All these compounds have an inhibitory effect on Cladosporium cucumerinum. Then, 4 coumarin compounds (8-11) are extracted from the leaves of *Murraya alate Drak*, all of which show excellent inhibitory activity against porphyromonas gingivalis. Natural coumarins 1-11 are described in Figure 36.

Coumarin derivatives synthesized by chemical ways also show excellent inhibitory effect on different bacteria, such as coumarin piperazine derivatives (Figure 37). They are synthesized using 4-hydroxycoumarin and epichlorohydrin as materials (Buckle et al., 1984). These compounds reveal good antibacterial activity against Gram-positive strains. When the hydrogen on the N of piperazine is replaced by 4-methoxyphenyl, their antibacterial activity is even better than penicillin. A series of coumarin esters have been synthesized (Figure 38). The antibacterial and antifungal activities of these compounds are investigated. The results prove that the antibacterial activities of some compounds are stronger than positive control ampicillin (Utreja et al., 2018).

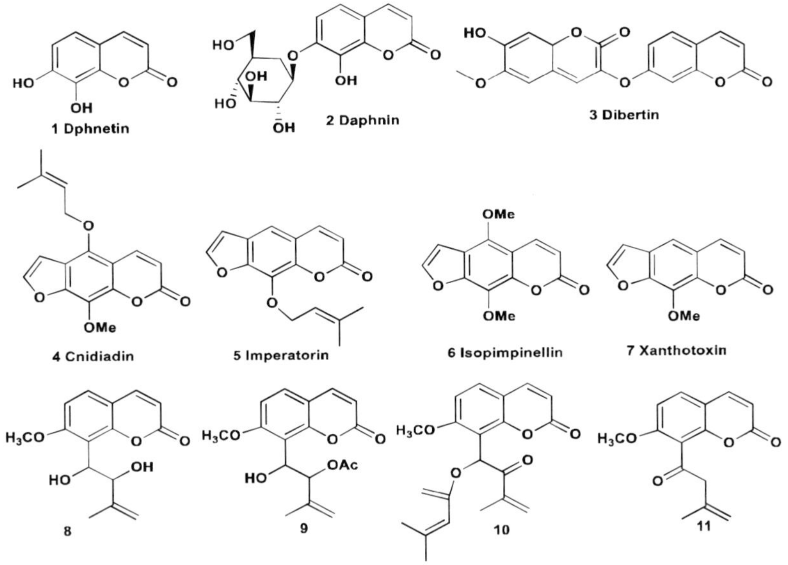

Figure 36. The structures of natural coumarins 1-11.

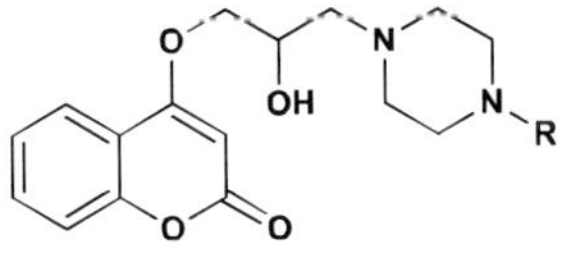

Figure 37. Structures of coumarin piperazine derivatives.

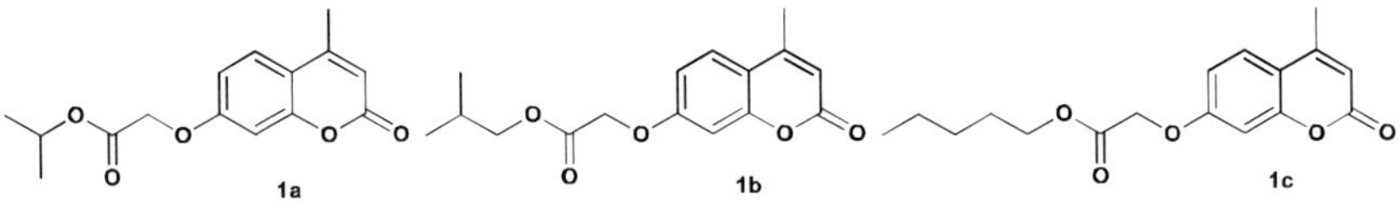

Figure 38. Structures of coumarin ester derivatives.

Other new coumarin aminophosphate derivatives are shown in Figure 39 (Ji et al., 2016). The results of chitin synthase activity and antimicrobial activity *in vitro* show that some compounds have certain selectivity in antibacterial activity. Coumarin derivatives with cyanopyridine are also synthesized by microwave method. These obtained compounds exhibit the same antibacterial activity as ampicillin (50 μg/mL). Coumarin derivatives with pyrazole groups (Figure 40) are reported (Chavan and Hosamani, 2018). The antibacterial and anti-inflammatory experiments on these compounds suggest that compounds (2b) and (2d) have good antibacterial effects. The minimum inhibitory concentration of compounds (2b) and (2d) against Gram-positive Staphylococcus aureus is 0.78 μg/mL and 1.562 ug/mL. Anti-inflammatory activity results confirm that compounds (2a) and (2c) are more effective than the standard drug acelofenac.

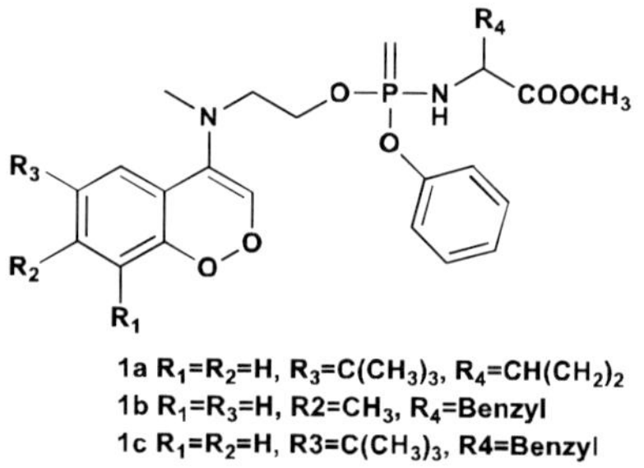

Figure 39. Structures of coumarin aminophosphate derivatives.

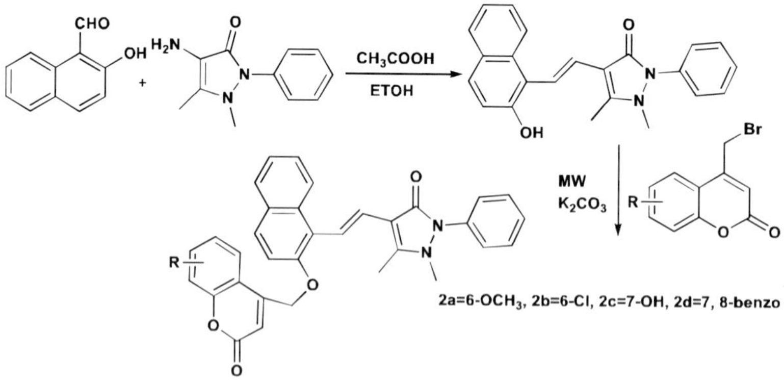

Figure 40. Structures of coumarin imidazole derivatives.

6.3. Antioxidant Activity

Four coumarin compounds are extracted from the bark of Poria cocos and show strong inhibitory effect on DPPH free radical, which is superior to positive control vitamin C (Al-Amiery et al., 2017). A variety of arylcoumarin derivatives substituted at the C-3 position are synthesized, and antioxidant tests show that the free radical scavenging activity is comparable to that of the positive control demethyldihydroguaiacic acid. Moreover, coumarin derivatives are synthesized from 2-bromoisobutyrate ethyl ester and p-hydroxyacetophenone. It is evaluated that the antioxidant activities of these compounds are better than that of water-soluble vitamin E as the positive control (Alshibl et al., 2020). Other six coumarins are extracted from Corydalis heterocarpa, which are (2′s)-columbianetin, (2′s)-columbianetin acetate, (2′s)-columbianetin glucoside, (2′s)-columbianetin 3′-propanoate, (2′s)-columbianetin 3′-sulfate, and (2′s)-columbianetin 3′-isopentanoate (Figure 41) (Li et al., 1989). They have good antioxidant activity. Excitedly, A novel coumarin, named 7-methoxy-8-(-3-methyl-2,3-epoxy-1-oxobutyl)-coumarin (Figure 42) (Sarker and Nahar, 2017), has been isolated from the fruit of moneil. Its oxidative activity is required to be further explored.

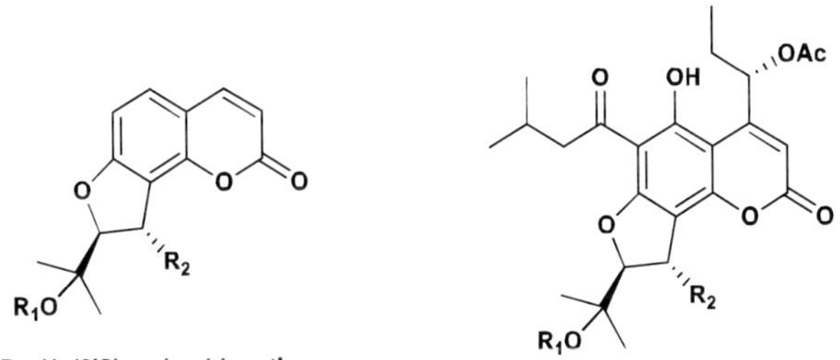

$R_1=R_2=H$, (2'S) -columbianetin
$R_1=AC$, $R_3=H$, (2'S) -columbianetin acetate
$R_1=$glucosyl, $R_2=H$, (2'S) -columbianetin glucoside

$R_1=$Propanoyl, (2'S) -columbianetin 3'S -propanoate
$R_1=SO_3H$, (2'S) -columbianetin 3'S -sulfate
$R_1=$(2'S) -columbianetin 3'S-isopentanoate

Figure 41. The structures of six coumarins extracted from Corydalis heterocarpa.

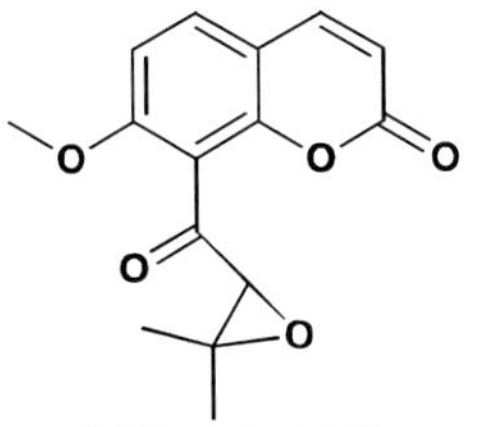

7-methoxy-8-(-3-methyl-2,3-epoxy-1-oxobutyl)-coumarin

Figure 42. The structure of 7-methoxy-8-(-3-methyl-2,3-epoxy-1-oxobutyl)-coumarin.

6.4. Antiviral and Anti-HIV Activity

Glycycoumarin and glycyrol (Figure 43) isolated from licorice can prevent hepatitis virus (Adianti et al., 2014). Peucedanone (21) has a certain inhibitory effect on the replication of HIV (Rajtar et al., 2012). Soulattrolide (22) isolated from teysmannii can effectively inhibit HIV-1-RT, and its IC_{50} value is 0.34 μmol/L, but only for AMV RT, DNA polymerase (α and β) and RNA polymerase (Pengsuparp et al., 1996). Peucedanone and soulattrolide are shown in Figure 44.

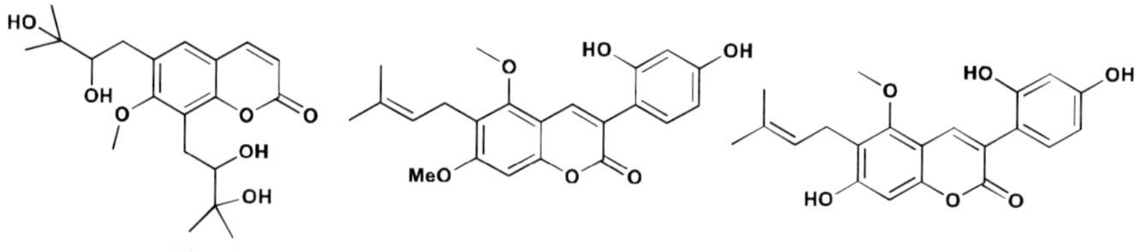

Figure 43. Natural coumarins from licorice.

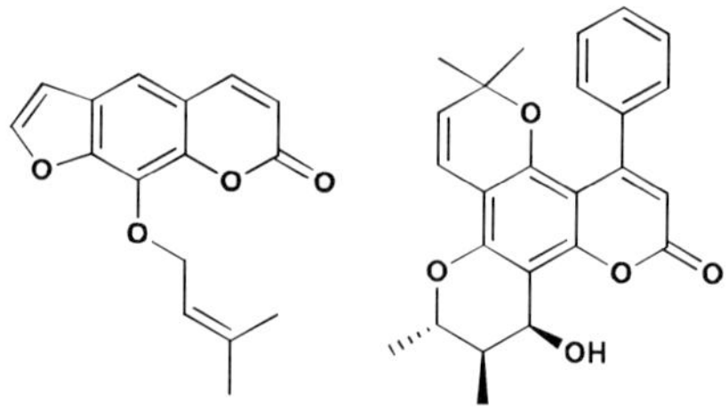

Figure 44. The structures of peucedanone and soulattrolide.

In this experiment, different derivatives of catechol and pyrogallol (Figure 45) are synthesized by the oxidation reaction of coumarin and dihydroxybenzoic acid in dimethyl sulfoxide at 25 °C (Taştemel et al., 2015). The novel synthesized derivatives show good antioxidant activity and inhibit influenza A/PR8/H1N1 virus. Coumarins with high oxidation may show antiviral activity based on the control of intracellular redox. Coumarin derivatives are obtained by structural modification of coumarins (Figure 46). Virus tests show that compounds 3a and 3c have good inhibition of human colon virus. EC_{50} values of them in HEK293 cells are 3.29 and 4.85 μmol/L, respectively. The EC_{50} values in RD cells are 2.5 and 3.98 μmol/L (Li et al., 2021a).

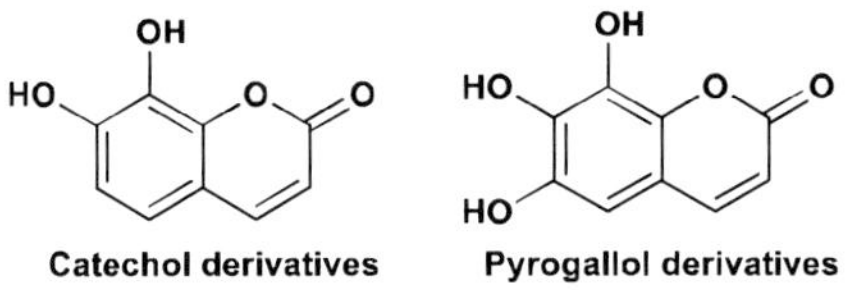
Catechol derivatives
Pyrogallol derivatives

Figure 45. Derivatives of catechol and pyrogallol.

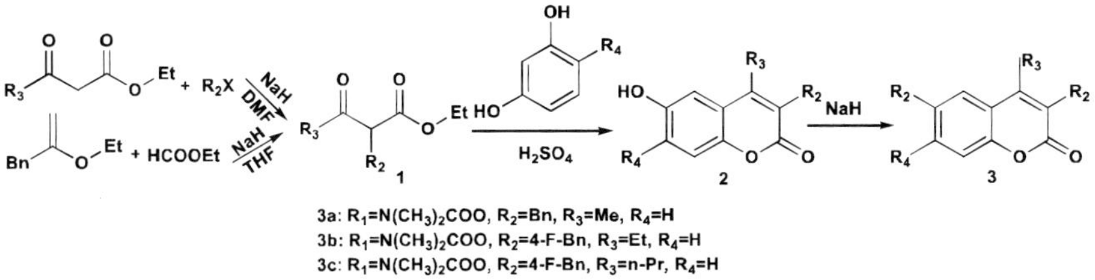
NaH
DMF
NaH
THF
H2SO4
NaH
3a: R1=N(CH3)2COO, R2=Bn, R3=Me, R4=H
3b: R1=N(CH3)2COO, R2=4-F-Bn, R3=Et, R4=H
3c: R1=N(CH3)2COO, R2=4-F-Bn, R3=n-Pr, R4=H

Figure 46. Coumarin derivatives.

Calanolide A (17) and calanolide B (18) extracted from Malaysian and *Calophyllum L.* have coumarin skeletons. They are highly selective and specific to HIV-1 reverse transcriptase, which can effectively inhibit the rapid replication of HIV and achieve the effect of treatment (Galinis et al., 1996). Calanolide A is currently in clinical phase II as an anti-HIV drug. In another experiment, the O atom on the B ring is replaced by the S atom to produce compound 19, which shows a stronger inhibition of HIV-1-RT than calanolide A (Zhan et al., 2013) At the same time, compound 20, the intermediate of compound 17, also shows good anti-HIV activity. Compounds 17-20 are illustrated in Figure 47 (Yu et al., 2003). Besides, triarylcoumarin is prepared by Suzuki-Miyauara coupling reaction using 4-methyl-6, 7-dihydroxy coumarin as raw material (Figure 48). In addition, the anti-HIV activity of the compounds is studied. Among them, compounds 3a and 3d show good anti-HIV activity with IC_{50} values of 4.57 and 13.20 umol/L.

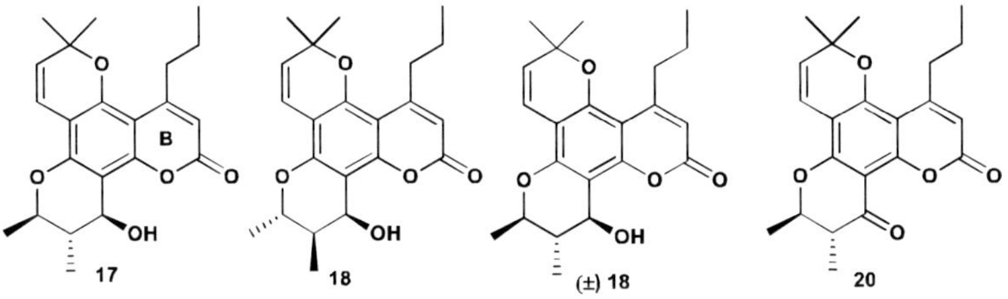
17
18
(±) 18
20

Figure 47. Synthetic coumarin derivatives with anti-HIV activity.

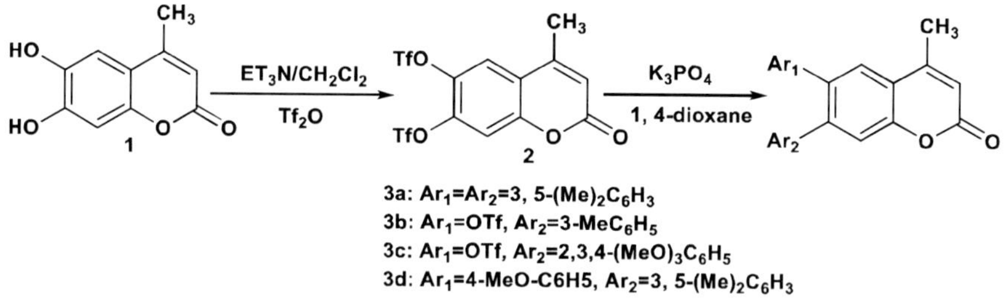

Figure 48. The synthetic route of triarylcoumarin.

6.5. Coumarin in the Treatment of Diabetes

Natural coumarins and their derivatives are proved to have anti-diabetic activity, with certain inhibitory ability on α-glucosidase (Islam et al., 2013). α-amylase and α-glucosidase are the main target enzymes of coumarin derivatives, which can inhibit hyperglycemia. High doses of coumarin can be toxic, such as carcinogenicity and hepatotoxicity. One or more pharmacophore groups are introduced into the existing coumarin molecules to improve bioactivities and reduce toxic and side effects through synergistic action (Riveiro et al., 2010).

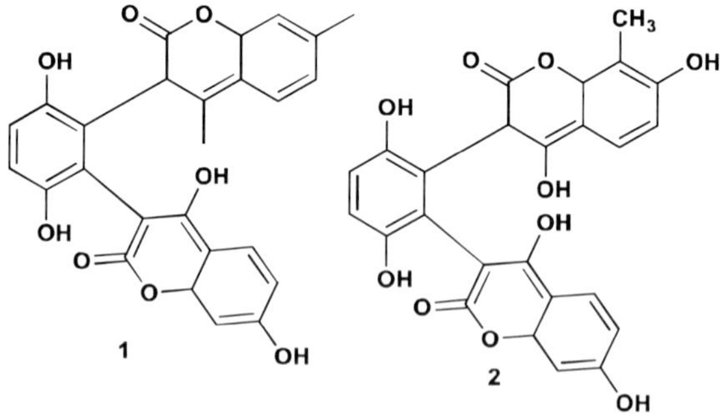

Figure 49. The structures of novel hydroxycoumarin derivatives.

Hydroxycoumarin derivatives are also synthesized, which show significant inhibitory activity against α-glucosidase (Singh et al., 2018). Here, compound 1 is the most potent α-glucosidase inhibitor with an IC_{50} value of 0.86 μM, which is about 14 times more active than genistein (IC_{50} = 12.36 μM). Compound 2 shows effective inhibitory activity with an IC_{50} value of

2.82 μM. When compound 1 interacts with α-glucosidase, the hydroxyl group exerts effect as both hydrogen bond donor and acceptor. The methoxy group of compound 2 acts only as a hydrogen bond acceptor. Therefore, compound 1 with excellent enzyme inhibition activity is more favorable to bind to α-glucosidase than compound 2. Both (Figure 49) reveal good activity against α-glucosidase and can be used as lead compounds to develop effective α-glucosidase inhibitors.

6.6. Insecticidal Activity

Due to the multiple pharmacological activities of coumarins, researchers have applied them as insect repellents. Accordingly, coumarin derivatives are synthesized by optimizing the structure of osthol as the lead compound, and then their insect repellent activity in goldfish is tested (Wang et al., 2012). The results confirm that compound 45 (Figure 50) exhibits the best insect repellent activity with EC_{50} value of 1.81 mg/L. In another experiment, benzene, acetone, methanol, and distilled water are used as solvents to extract 4 different extracts from the seeds of *Angelica dahurica* (Kwon et al., 1997). GC/MS and HPLC are used to detect the components. The results suggest that 50–56% of them are furancoumarins. Then the acute and chronic toxicity of these extracts on the antifeedant and growth inhibition of the larvae of Spodoptera littoralis are detected. The data reveal that these extracts have antifeedant and growth inhibiting effects on the larvae of Spodoptera littoralis. Here, the substance extracted with benzene possess the strongest antifeedant and growth inhibition effects on the larvae of Spodoptera littoralis, with EC_{50} values of 2.4 μg/g and 0.31 μg/g respectively. The main components of benzene extract include zanthoxylin, bergamot vinegar, imperatorin, shanchaogenin, isoimperatorin, etc.

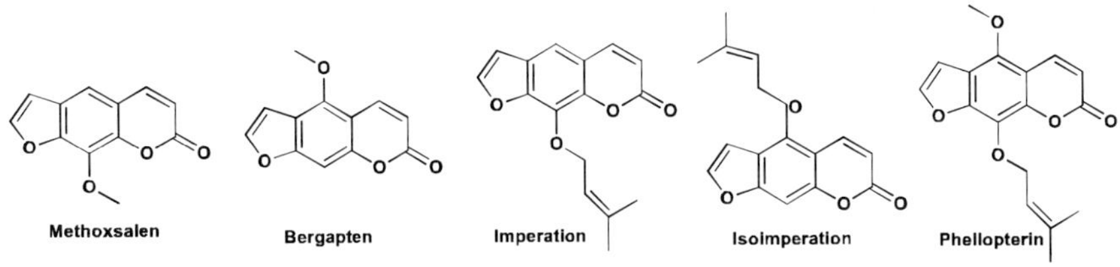

Figure 50. Coumarin derivatives by modifying the structure of osthol as the lead compound.

Conclusion and Prospects

Coumarins and their derivatives are more widely used in medicines. They are important compounds in natural pharmaceutical chemistry, which have obvious bioactivities of anticoagulation, anti-tumor and anti-human immunodeficiency virus. The structurally modified parent ring of coumarins has important value in the research and development of new drugs. In addition, they have good activities in inhibiting tyrosinase activity. This is helpful to find new food preservative and tyrosinase inhibitors. Due to the variety of natural coumarin derivatives, their antibacterial activities and mechanisms are not the same. Further study on the targets and antibacterial mechanism of coumarins will provide a new strategy for the treatment of drug-resistant bacterial infections and a new direction for the development of antibacterial drugs.

With the application of coumarins and their derivatives in medicine, pesticide, biology and materials, the reports about their synthesis methods are increasing year by year. These compounds with good bioactivities can be obtained by structural modification of coumarin parent nucleus, which has good application prospect. Nowadays, these compounds are mainly synthesized by traditional methods, Pechmann reaction method and o-hydroxyacetophenone method, while microbial synthesis methods are also gradually developed. Since coumarin inhibitors have been isolated from natural plants, various coumarin derivatives have been designed and synthesized, and pharmacological activity tests and structure-activity relationships have been conducted. It is found that the size of substituents, steric hindrance, substitution location and lipophile are the key factors affecting the inhibitory activity. Some lead compounds with good activity, high selectivity, and low cytotoxicity are obtained. With the development of high-throughput drug screening and computer-aided drug design ways, the research and development of anti-cancer coumarins with targeting multiple signal pathways or targets will be a hotspot for the development of anticancer drugs in the future.

A variety of fluorescent brighteners with different fluorescence properties can be synthesized by introducing various substituents into the coumarin ring, which can be used to modify and improve the structures of coumarins. Their derivatives are expected to be widely used in rapid and accurate ion detection, environmental detection, and water sample analysis. Meanwhile, coumarins and their derivatives have good optical properties used in fluorescent probes, such as high fluorescence quantum yield, large Stokes displacement, good solubility, and strong optical stability. Recently, fluorescent probes based on coumarins have been used in medicine, heavy metal ions, sulfur-containing compounds, reactive oxygen species, proteins, and cell structures. At present, coumarin fluorescent probes mainly have the problems involved in high synthesis cost, complex synthesis route, and environmental pollution. Different fluorescence probes with high selectivity, high sensitivity, simple operation, good solubility, non-toxic or low toxicity are designed by searching for new recognition sites and action mechanisms using coumarin parent nucleus as fluorescence emitters.

Coumarins as one of the best fluorophores in fluorescent probes exhibit high fluorescence intensity, good solubility and cell permeability, easy synthesis, and modification. Fluorescent probes are important for basic biological researches, diagnosis, and treatment of diseases, and provide "visualization" possibilities. However, there are more challenges to explore novel fluorescent probes. First, it is still challenging to develop probes with truly high selectivity and sensitivity. For example, the structures of Cys, Hcy and GSH are highly similar, and the concentration of GSH is at the level of millimole, while the concentration of Cys and Hcy is only at the micromole level, so it is difficult to detect Cys/Hcy specifically. Secondly, the environment in organisms is complex, which puts forward higher requirements for the study of fluorescence probe mechanism. Notably, coumarins have a large dipole moment and photosensitive properties, which makes the liquid crystal small molecules and polymers with coumarin groups have a wide liquid crystal range, good liquid crystal stability, as well as outstanding photoelectric properties and strong fluorescence. The photodimers of coumarin liquid crystal exert great effect in the orientation of liquid crystal and has become new functional materials. Coumarin liquid crystal compounds have great potential applications in fluorescent materials and laser dyes, nonlinear optical materials, photoresists, pharmaceutical intermediates, and luminescent materials. Considering that, the design and synthesis of more coumarin liquid crystal compounds is extremely important for the development of new functional properties of these compounds, which are worthy of continuous exploration and research by chemists.

References

Adianti, M., Aoki, C., Komoto, M., Deng, L., Shoji, I., Wahyuni, T.S., Lusida, M.I., Fuchino, H., Kawahara, N., and Hotta, H. (2014). Anti-hepatitis C virus compounds obtained from *Glycyrrhiza uralensis* and other *Glycyrrhiza species*. *Microbiology and Immunology* 58, 180-187.

Akhtar, J., Khan, A.A., Ali, Z., Haider, R., and Yar, M.S. (2017). Structure-activity relationship (SAR) study and design strategies of nitrogen-containing heterocyclic moieties for their anticancer activities. *European Journal of Medicinal Chemistry* 125, 143-189.

Al-Amiery, A.A., Saour, K.Y., A-Duhaidahawi, D.L., Al-Majedy, Y.K., Kadhum, A.A., and Mohamad, A.B. (2017). Comparative molecular modelling studies of coumarin derivatives as potential antioxidant agents. *Free Radicals and Antioxidants* 7, 31-35.

Al-Majedy, Y., Al-Amiery, A., Kadhum, A.A., and BakarMohamad, A. (2017). Antioxidant activity of coumarins. *Systematic Reviews in Pharmacy* 8, 24-30.

Alamzeb, M., Ali, S., Khan, A.A., Igoli, J.O., Ferro, V.A., Gray, A.I., and Khan, M.R. (2015). A new ceramide along with eight known compounds from the roots of *Artemisia incisa* pamp. *Records of Natural Products* 9, 297-304.

Alshibl, H.M., Al-Abdullah, E.S., Haiba, M.E., Alkahtani, H.M., Awad, G.E., Mahmoud, A.H., Ibrahim, B.M., Bari, A., and Villinger, A. (2020). Synthesis and evaluation of new coumarin derivatives as antioxidant, antimicrobial, and anti-inflammatory agents. *Molecules* 25, 3251.

Aslam, K., Khosa, M.K., Jahan, N., and Nosheen, S. (2010). Short communication: Synthesis and applications of coumarin. *Pakistan Journal of Pharmaceutical Sciences* 23, 449-454.

Aydin, Z., Keskinates, M., Yilmaz, B., Durmaz, M., and Bayrakci, M. (2022). A rapid responsive coumarin-naphthalene derivative for the detection of cyanide ions in cell culture. *Analytical Biochemistry*, 114798.

Azagarsamy, M.A., McKinnon, D.D., Alge, D.L., and Anseth, K.S. (2014). Coumarin-based photodegradable hydrogel: Design, synthesis, gelation, and degradation kinetics. *ACS Macro Letters* 3, 515-519.

Babaei, E., Küçükkılınç, T.T., Jalili-Baleh, L., Nadri, H., Öz, E., Forootanfar, H., Hosseinzadeh, E., Akbari, T., Ardestani, M.S., and Firoozpour, L. (2022). Novel coumarin–pyridine hybrids as potent multi-target directed ligands aiming at symptoms of alzheimer's disease. *Frontiers in Chemistry* 10, 895483.

Bana, E., Sibille, E., Valente, S., Cerella, C., Chaimbault, P., Kirsch, G., Dicato, M., Diederich, M., and Bagrel, D. (2015). A novel coumarin-quinone derivative SV37 inhibits CDC25 phosphatases, impairs proliferation, and induces cell death. *Molecular Carcinogenesis* 54, 229-241.

Bansal, Y., Sethi, P., and Bansal, G. (2013). Coumarin: A potential nucleus for anti-inflammatory molecules. *Medicinal Chemistry Research* 22, 3049-3060.

Barriga-Gonzalez, G., and Munoz-Espinoza, J. (2022). Thermodynamic and reactivity aspect of B-cyclodextrine inclusion complexes with coumarin derivatives. *Journal of the Chilean Chemical Society* 67, 5514-5520.

Bonelli, J., Rovira, A., Marchan, V., Bonelli, J., Torres, O., Cusco, C., Rocas, J., Ortega-Forte, E., Ruiz, J., and Bosch, M. (2022). Improving photodynamic therapy anticancer activity of a mitochondria-targeted coumarin photosensitizer using a polyurethane-polyurea hybrid nanocarrier. *Biomacromolecules* 23, 2900-2913.

Buckle, D.R., Outred, D.J., Smith, H., and Spicer, B.A. (1984). N-benzylpiperazino derivatives of 3-nitro-4-hydroxycoumarin with H1 antihistamine and mast cell stabilizing properties. *Journal of Medicinal Chemistry* 27, 1452-1457.

Cao, J., Wu, Q., Chang, X., Chu, H., Zhang, H., Fang, X., and Chen, F. (2022). Ratiometric detection and imaging of endogenous alkaline phosphatase activity by fluorescein-coumarin- based fluorescence probe. *Spectrochimica Acta Part A: Molecular and Biomolecular Spectroscopy*, 121615.

Chatterjee, R., Santra, S., Zyryanov, G.V., and Majee, A. (2019). Vinylation of carbonyl oxygen in 4-hydroxycoumarin: synthesis of heteroarylated vinyl ethers. *Synthesis* 51, 2371-2378.

Chaudhari, D.D. (1983). Heterogeneous catalysis by solid superacid: Nafion-H-catalyzed von Pechmann condensation. *Chemistry & Industry*, 568-569.

Chavan, R.R., and Hosamani, K.M. (2018). Microwave-assisted synthesis, computational studies and antibacterial/anti-inflammatory activities of compounds based on coumarin-pyrazole hybrid. *Royal Society Open Science* 5, 172435.

Cheke, R.S., Ambhore, J.P., Patel, H.M., Ansari, I.A., Patil, V.M., Shinde, S.D., Kadri, A., Kadri, A., Snoussi, M., Adnan, M., et al. (2022). Molecular insights into coumarin analogues as antimicrobial agents: Recent developments in drug discovery. *Antibiotics* 11, 566.

Chen, H., and Walsh, C.T. (2001). Coumarin formation in novobiocin biosynthesis: β-hydroxylation of the aminoacyl enzyme tyrosyl-S-NovH by a cytochrome P450 NovI. *Chemistry & Biology* 8, 301-312.

Chen, Y., Fan, G., Zhang, Q., Wu, H., and Wu, Y. (2007). Fingerprint analysis of the fruits of Cnidium monnieri extract by high-performance liquid chromatography–diode array detection–electrospray ionization tandem mass spectrometry. *Journal of Pharmaceutical and Biomedical Analysis* 43, 926-936.

Cho, N., Kikuzato, K., Futamura, Y., Shimizu, T., Hayase, H., Osada, H., Kamisaka, K., Takaya, D., Yuki, H., Honma, T., et al. (2022). New antimalarials identified by a cell-based phenotypic approach: Structure-activity relationships of 2,3,4,9-tetrahydro-1H-β-carboline derivatives possessing a 2-((coumarin-5-yl)oxy)alkanoyl moiety. *Bioorganic & Medicinal Chemistry* 66, 116830.

Choi, T.J., Song, J., Park, H.J., Kang, S.S., and Lee, S.K. (2022). Anti-inflammatory activity of glabralactone, a coumarin compound from angelica sinensis, via suppression of TRIF-dependent IRF-3 signaling and NF-κB pathways. *Mediators of Inflammation* 2022, 5985255.

Coulibaly, S., Teki, D.S.-E.K., Coulibali, S., Ablo, E., and Adjou, A. (2022). Synthesis of 6-phenylethynyl substituted coumarins via a Sonogashira coupling. *Pharmaceutical Chemistry Journal* 9, 12-23.

Debbabi, K.F., Al-Harbi, S.A., Al-Saidi, H.M., Bashandy, M.S., Debbabi, K.F., Aljuhani, E.H., Abd, E.-G.S.M., and Bashandy, M.S. (2016). Study of reactivity of cyanoacetohydrazonoethyl-N-ethyl-N-methyl benzenesulfonamide: Preparation of novel anticancer and antimicrobial active heterocyclic benzenesulfonamide derivatives and their molecular docking against dihydrofolate reductase. *Journal of Enzyme Inhibition and Medicinal Chemistry* 31, 7-19.

Detsi, A., Kontogiorgis, C., and Hadjipavlou-Litina, D. (2017). Coumarin derivatives: An updated patent review (2015-2016). *Expert Opinion on Therapeutic Patents* 27, 1201-1226.

Farhat, N., Ali, A., Gupta, D., Khan, A.U., and Waheed, M. (2022). Chemically synthesised flavone and coumarin based isoxazole derivatives as broad spectrum inhibitors of serine β-lactamases and metallo-β-lactamases: A computational, biophysical and biochemical study. *Journal of Biomolecula Structure & Dynamics*, 1-11.

Feng, Y.-M., Qi, P.-Y., Xiao, W.-L., Zhang, T.-H., Zhou, X., Liu, L.-W., and Yang, S. (2022). Fabrication of isopropanolamine-decorated coumarin derivatives as novel quorum sensing inhibitors to suppress plant bacterial disease. *Journal of Agricultural and Food Chemistry* 70, 6037-6049.

Ferreira, A.R., de, S.D.P., Alves, D.d.N., de, C.R.D., and Perez-Castillo, Y. (2022). Synthesis of coumarin and homoisoflavonoid derivatives and analogs: The search for new antifungal agents. *Pharmaceuticals* 15, 712.

Franchin, M., Luiz, R.P., da, S.P.D., Leal, S.R., Damasceno, L.E.A., Morelo, D.F.C., Cunha, F.Q., Alves-Filho, J.C., Cesar, P.M., Napimoga, M.H., et al. (2022). Cinnamoyloxy-mammeisin, a coumarin from propolis of stingless bees, attenuates Th17 cell differentiation and autoimmune inflammation via STAT3 inhibition. *European Journal of Pharmacology* 929, 175127.

Freitas, N.d.S.W.J., Sernizon, G.N., Parreiras, M.M.A., Freitas, N.d.S.W.J., Margotto, B.C., Parreiras, M.M.A., Costa, V.C., Silva, M.P.T., Fonseca, M.A., Sousa, V.M., et al. (2022). Factors associated with nonadherence to the use of coumarin derivatives or direct oral anticoagulants: A systematic review of observational studies. *British Journal of Clinical Pharmacology*.

Fu, Z., Zhang, L., Hang, S., Li, N., Sun, X., Wang, Z., Wu, W., Guo, R., Wang, S., Sheng, R., et al. (2022). Synthesis of coumarin derivatives: A new class of coumarin-based G protein-coupled receptor activators and inhibitors. *Polymers* 14, 2021.

Fuentes-Aguilar, A., Merino-Montiel, P., Montiel-Smith, S., Meza-Reyes, S., Vega-Baez, J.L., Puerta, A., Fernandes, M.X., Padron, J.M., Petreni, A., Nocentini, A., et al. (2022). 2-Aminobenzoxazole-appended coumarins as potent and selective inhibitors of tumour-associated carbonic anhydrases. *Journal of Enzyme Inhibition and Medicinal Chemistry* 37, 168-177.

Galinis, D.L., Fuller, R.W., McKee, T.C., Cardellina, J.H., Gulakowski, R.J., McMahon, J.B., and Boyd, M.R. (1996). Structure–activity modifications of the HIV-1 inhibitors (+)-calanolide A and (−)-calanolide B. *Journal of Medicinal Chemistry* 39, 4507-4510.

Gao, R., Liu, X., Feng, J., Han, L., Xu, J., and Kan, C. (2022). Synthesis and application of a novel polyurethane nanoemulsion bearing coumarin derivative as a "turn-on" fluorescence sensor toward Hg(II). *Spectrochimica Acta Part A: Molecular and Biomolecular Spectroscopy* 281, 121612.

Ghazouani, L., Elmufti, A., Feriani, A., Baaziz, I., Khdhiri, E., Ammar, H., Zarei, A., Ramazani, A., Hajji, R., Abid, M., et al. (2022). A novel synthetized sulfonylhydrazone coumarin (E)-4-methyl-N'-(1-(3-oxo-3H-benzo[f]chromen-2-yl)ethylidene)benzenesulfonohydrazide protect against isoproterenol induced myocardial infarction in rats by attenuating oxidative damage, biological changes, and electrocardiogram. *Clinical and Experimental Pharmacology and Physiology*, Doi: 10.1111/1440-1681.13690.

Ghosh, A., Ranjan, N., Jiang, L., Ansari, A.H., Degyatoreva, N., Ahluwalia, S., Arya, D.P., and Maiti, S. (2022). Fine-tuning miR-21 expression and inhibition of EMT in breast cancer cells using aromatic-neomycin derivatives. *Molecular Therapy-Nucleic Acids* 27, 685-698.

Gopalakrishnan, S., Viswanathan, K., Priya, S.V., Mabel, J.H., Palanichamy, M., and Murugesan, V. (2009). Synthesis of 7-hydroxy-4-methyl coumarin over Lewis acid metal ion-exchanged ZAPO-5 molecular sieves. *Microporous and Mesoporous Materials* 118, 523-530.

Hanessian, S., Giguère, A., Grzyb, J., Maianti, J.P., Saavedra, O.M., Aggen, J.B., Linsell, M.S., Goldblum, A.A., Hildebrandt, D.J., and Kane, T.R. (2011). Toward overcoming Staphylococcus aureus aminoglycoside resistance mechanisms with a functionally designed neomycin analogue. *ACS Medicinal Chemistry Letters* 2, 924-928.

Heravi, M.M., Khaghaninejad, S., and Mostofi, M. (2014). Pechmann reaction in the synthesis of coumarin derivatives. *Advances in Heterocyclic Chemistry* 1-50.

Hosseinzadeh, Z., Ramazani, A., and Razzaghi-Asl, N. (2019). Plants of the genus heracleum as a source of coumarin and furanocoumarin. *Journal of Chemical Reviews* 1, 78-98.

Houshmand, A., Liu, D.Y., Zhou, W., Linington, R.G., Warren, J.J., Walsby, C.J., Heroux, D., and Bally, M. (2022). Ferrocene-appended anthraquinone and coumarin as redox-active cytotoxins. *Dalton Transactions* Doi:10.1039/D2DT01251K.

Hu, C.-M., Luo, Y.-X., Wang, W.-J., Li, J.-P., Li, M.-Y., Zhang, Y.-F., Xiao, D., Lu, L., Xiong, Z., Feng, N., et al. (2022). Synthesis and evaluation of coumarin-chalcone derivatives as α-glucosidase inhibitors. *Frontiers in Chemistry* 10, 926543.

Huang, L., Feng, Z.-L., Wang, Y.-T., and Lin, L.-G. (2017). Anticancer carbazole alkaloids and coumarins from Clausena plants: a review. *Chinese Journal of Natural Medicines* 15, 881-888.

Huang, X., and Liu, J. (2018). Synthesis and anticancer activities of novel pyranocoumarin fused pyrimidine based on cyanoenamine. *Chinese Journal of Organic Chemistry* 38, 1233-1241.

Ibrahim, H.S., Abdelrahman, M.A., Nocentini, A., Bua, S., Abdel-Aziz, H.A., Supuran, C.T., Abou-Seri, S.M., and Eldehna, W.M. (2022a). Insights into the effect of elaborating coumarin-based aryl enaminones with sulfonamide or carboxylic acid functionality on carbonic anhydrase inhibitory potency and selectivity. *Bioorganic Chemistry* 126, 105888.

Ibrahim, M.E.-D., El-Zend, M.A., Tantawy, M.A., and Barakat, L.A.A. (2022b). Synthesis and cytotoxicity screening of some synthesized coumarin and aza-coumarin derivatives as anticancer agents. *Russian Journal of Bioorganic Chemistry* 48, 380-390.

Islam, N., Jung, H.A., Sohn, H.S., Kim, H.M., and Choi, J.S. (2013). Potent α-glucosidase and protein tyrosine phosphatase 1B inhibitors from Artemisia capillaris. *Archives of Pharmacal Research* 36, 542-552.

Iyer, D., and Patil, U. (2014). Evaluation of antihyperlipidemic and antitumor activities of isolated coumarins from Salvadora indica. *Pharmaceutical Biology* 52, 78-85.

Jana, A., Ali, D., Bhaumick, P., and Choudhury, L.H. (2022). Sc(OTf)$_3$-mediated one-pot synthesis of coumarin-fused furans: A thiol-dependent reaction for the easy access of 2-phenyl-4H-furo[3,2-c]chromen-4-ones. *Journal of Organic Chemistry* 87, 7763-7777.

Ji, Q., Ge, Z., Ge, Z., Chen, K., Wu, H., Liu, X., Huang, Y., Yuan, L., Yang, X., and Liao, F. (2016). Synthesis and biological evaluation of novel phosphoramidate derivatives of coumarin as chitin synthase inhibitors and antifungal agents. *European Journal of Medicinal Chemistry* 108, 166-176.

Karatas, M.O., Alici, B., Passarelli, V., Ozdemir, I., Perez-Torrente, J.J., and Castarlenas, R. (2021). Iridium(I) complexes bearing hemilabile coumarin-functionalized N-heterocyclic carbene ligands with application as alkyne hydrosilylation catalysts. *Dalton Transactions* 50, 11206-11215.

Karatas, M.O., Tekin, S., Alici, B., and Sandal, S. (2019). Cytotoxic effects of coumarin substituted benzimidazolium salts against human prostate and ovarian cancer cells. *Journal of Chemical Sciences* 131, 1-12.

Karcz, D., Starzak, K., Ciszkowicz, E., Lecka-Szlachta, K., Kaminski, D., Creaven, B., Milos, A., Jenkins, H., Slusarczyk, L., and Matwijczuk, A. (2022). Design, Spectroscopy, and assessment of cholinesterase inhibition and antimicrobial activities of novel coumarin-thiadiazole hybrids. *International Journal of Molecular Sciences* 23.

Kaur, P., Gill, R.K., Singh, G., and Bariwal, J. (2016). Synthesis, cytotoxic evaluation, and in silico studies of 4-substituted coumarins. *Journal of Heterocyclic Chemistry* 53, 1519-1527.

Kosenko, I., Laskova, J., Kozlova, A., Semioshkin, A., and Bregadze, V.I. (2020). Synthesis of coumarins modified with cobalt bis (1,2-dicarbolide) and closo-dodecaborate boron clusters. *Journal of Organometallic Chemistry* 921, 121370.

Kowalska, E., Artelska, A., and Albrecht, A. (2022). Visible light-driven reductive azaarylation of coumarin-3-carboxylic acids. *Journal of Organic Chemistry* Doi:10.1021/acs.joc.2c00683.

Kuchlyan, J., Banik, D., Kundu, N., Roy, A., and Sarkar, N. (2014). Interaction of fluorescence dyes with 5-fluorouracil: A photoinduced electron transfer study in bulk and biologically relevant water. *Chemical Physics Letters* 613, 115-121.

Kumar, A., Kumar, P., and Pai, A. (2021). Synthesis, docking and antimicrobial activity of some new coumarin incorporated thiazole derivatives. *Journal of Pharmaceutical Research International* 33, 332-340.

Kusumoto, T., Inaniwa, T., Mizushima, K., Sato, S., Hojo, S., Kitamura, H., Konishi, T., and Kodaira, S. (2022). Radiation chemical yields of 7-hydroxy-coumarin-3-carboxylic acid for proton- and carbon-ion beams at ultra-high dose rates: Potential roles in FLASH effects. *Radiation Research* Doi:10.1667/RADE-21-00.230.1.

Kwon, Y.-S., Kobayashi, A., Kajiyama, S.-I., Kawazu, K., Kanzaki, H., and Kim, C.-M. (1997). Antimicrobial constituents of *Angelica dahurica* roots. *Phytochemistry* 44, 887-889.

Lamani, S.S., Kotresh, O., A Phaniband, M.S., and C Kadakol, J. (2009). Synthesis, characterization and antimicrobial properties of schiff bases derived from condensation of 8-formyl-7-hydroxy-4-methylcoumarin and substituted triazole derivatives. *Journal of Chemistry* 6, S239-S246.

Langat, M.K., Randrianavelojosia, M., Langat, L.C., Plant, N., and Mulholland, D.A. (2014). Langaside, a novel secoiridolactone glycoside derivative from *Tachiadenus longiflorus* Griseb. (Gentianaceae) formed by a [2+2] cycloaddition reaction. *Phytochemistry Letters* 10, cxviii-cxxii.

Larbat, R., Hehn, A., Hans, J., Schneider, S., Jugde, H., Schneider, B., Matern, U., and Bourgaud, F. (2009). Isolation and functional characterization of CYP71AJ4 encoding for the first P450 monooxygenase of angular furanocoumarin biosynthesis. *Journal of Biological Chemistry* 284, 4776-4785.

Li, J.-M., Huang, T.-T., Xu, N., Xu, X.-Y., and Yu, H.-C. (2022a). The facial and short method for the synthesis of Z/E-isomers of acetate of 12-tetradecene-1-ol components of the sex pheromone ostrinia furnacalis. *Chemistry of Natural Compounds* 58, 520-523.

Li, J.-Y., Yin, L.-P., Jiang, X.-M., Guo, K., Zhang, C., Wang, Q., Wang, L., Yin, L.-P., Wang, Q., Wang, L., et al. (2022b). A racemic naphthyl-coumarin-based probe for quantitative enantiomeric excess analysis of amino acids and enantioselective sensing of amines and amino alcohols. *ChemistryOpen* 11, e202200088.

Li, J., Feng, L., Liu, L., Wang, F., Ouyang, L., Zhang, L., Hu, X., and Wang, G. (2021a). Recent advances in the design and discovery of synthetic tyrosinase inhibitors. *European Journal of Medicinal Chemistry* 224, 113744.

Li, L., and Seeram, N.P. (2010). Maple syrup phytochemicals include lignans, coumarins, a stilbene, and other previously unreported antioxidant phenolic compounds. *Journal of Agricultural and Food Chemistry* 58, 11673-11679.

Li, N., Liu, X., Zhang, M., Zhang, Z., Zhang, B., Wang, X., Wang, J., Tu, P., Shi, S.-P., Li, N., et al. (2022c). Characterization of a coumarin C-/O-prenyltransferase and a quinolone C-prenyltransferase from *Murraya exotica*. *Organic & Biomolecular Chemistry* 20, 5535-5542.

Li, R., He, Y., Chiao, M., Xu, Y., Zhang, Q., Meng, J., Gu, Y., and Ge, L. (1989). Studies of the active constituents of the Chinese drug "duhuo" *Angelica pubescents*. *Acta Pharmaceutica Sinica* 24, 546-551.

Li, Y., Tang, Q., Xie, Y., He, D., Yang, K., and Zheng, L. (2021b). Synthesis of mitochondria-targeted coumarin-3-carboxamide fluorescent derivatives: Inhibiting mitochondrial TrxR2 and cell proliferation on breast cancer cells. *Bioorganic & Medicinal Chemistry Letters* 33, 127750.

Lingaraju, G.S., Balaji, K.S., Jayarama, S., Anil, S.M., Kiran, K.R., and Sadashiva, M.P. (2018). Synthesis of new coumarin tethered isoxazolines as potential anticancer agents. *Bioorganic & Medicinal Chemistry Letters* 28, 3606-3612.

Liu, B., Raeth, T., Beuerle, T., and Beerhues, L. (2010). A novel 4-hydroxycoumarin biosynthetic pathway. *Plant Molecular Biology* 72, 17-25.

Liu, F.-T., Chen, Y.-S., Yu, H.-Y., Li, N., Miao, J.-Y., and Zhao, B.-X. (2022a). A quinoline-coumarin near-infrared ratiometric fluorescent probe for detection of sulfur dioxide derivatives. *Analytica Chimica Acta* 1211, 339908.

Liu, X., Xu, C., Fan, F., Chen, P., Wang, Y., Zhu, M., and Li, D. (2022b). An ICT-based coumarin fluorescent probe for the detection of hydrazine and its application in environmental water samples and organisms. *Frontiers in Bioengineering and Biotechnology* 10, 937489.

Llanos, M.A., Enrique, N., Sbaraglini, M.L., Garofalo, F.M., Talevi, A., Gavernet, L., and Martín, P. (2022). Structure-based virtual screening identifies novobiocin, montelukast, and cinnarizine as TRPV1 modulators with anticonvulsant activity in vivo. *Journal of Chemical Information and Modeling*, 3008-3022.

Long, Q., Feng, L., Li, Y., Zuo, T., Chang, L., Zhang, Z., and Xu, P. (2022). Time-resolved quantitative phosphoproteomics reveals cellular responses induced by caffeine and coumarin. *Toxicology and Applied Pharmacology* 449, 116115.

Lu, P.-H., Liao, T.-H., Lu, P.-H., Chen, Y.-H., Hsu, Y.-L., Chern, C.-Y., Kuo, C.-Y., Tsai, F.-M., Chan, C.-C., and Wang, L.-K. (2022). Coumarin derivatives inhibit ADP-induced platelet activation and aggregation. *Molecules* 27, 4054.

Luo, D., Zhang, X., Li, X., Zhen, Y.-Y., Zeng, X., Xiong, Z., Zhang, Y., Li, H., Zhang, Y., and Li, H. (2022). Responsive fluorescent coumarin-cinnamic acid conjugates for α-glucosidase detection. *Frontiers in Chemistry* 10, 927624.

Luo, G., Muyaba, M., Lyu, W., Tang, Z., Zhao, R., Xu, Q., You, Q., and Xiang, H. (2017). Design, synthesis and biological evaluation of novel 3-substituted 4-anilino-coumarin derivatives as antitumor agents. *Bioorganic & Medicinal Chemistry Letters* 27, 867-874.

Malankar, G.S., Shelar, D.S., Manjare, S.T., and Butcher, R.J. (2022). Synthesis and single crystal X-ray study of phenylselenyl embedded coumarin-based sensors for selective detection of superoxide. *Dalton Transactions* 51, 10518-10526.

Marriott, K.-S.C., Bartee, R., Morrison, A.Z., Stewart, L., and Wesby, J. (2012). Expedited synthesis of benzofuran-2-carboxylic acids via microwave-assisted Perkin rearrangement reaction. *Tetrahedron Letters* 53, 3319-3321.

Moghadam, F.S., Seifinoferest, B., Larijani, B., Mahdavi, M., and Gholamhosseyni, M. (2022). Modern metal-catalyzed and organocatalytic methods for synthesis of coumarin derivatives: A review. *Organic & Biomolecular Chemistry* 20, 4846-4883.

Mohammed, A.E., Alotaibi, M.O., and Elobeid, M. (2022). Interactive influence of elevated CO_2 and arbuscular mycorrhizal fungi on sucrose and coumarin metabolism in Ammi majus. *Plant Physiology and Biochemistry* 185, 45-54.

Mustafa, Y.F., Bashir, M.K., and Oglah, M.K. (2020). Original and innovative advances in the synthetic schemes of coumarin-based derivatives: A review. *Systematic Reviews in Pharmacy* 11, 598-612.

Naik, V.G., Hiremath, S.D., Thakuri, A., Banerjee, M., Chatterjee, A., Hemmadi, V., and Biswas, M. (2022). A coumarin coupled tetraphenylethylene based multi-targeted AIEgen for cyanide ion and nitro explosive detection, and cellular imaging. *Analyst* 147, 2997-3006.

Niwa, T., Sasaki, S., Yamamoto, Y., and Tanaka, M. (2022). Effect of human cytochrome P450 2D6 polymorphism on progesterone hydroxylation. *European Journal of Drug Metabolism and Pharmacokinetics* Doi:10.1007/s13318-022-00784-7.

Onder, F.C., Sahin, K., Senturk, M., Durdagi, S., and Ay, M. (2022). Identifying highly effective coumarin-based novel cholinesterase inhibitors by in silico and *in vitro* studies. *Journal of Molecular Graphics and Modelling* 115, 108210.

Parvej, H., Begum, S., Dalui, R., Paul, S., Mondal, B., Sardar, S., Maiti, G., Halder, U.C., and Sepay, N. (2022). Coumarin derivatives inhibit the aggregation of β-lactoglobulin. *RSC Advances* 12, 17020-17028.

Patil, S.M., Martiz, R.M., Parameswaran, S., Ramu, R., Satish, A.M., Shbeer, A.M., Ageel, M., Al-Ghorbani, M., Al-Ghorbani, M., and Ranganatha, L.T. (2022). Discovery of novel coumarin derivatives as potential dual inhibitors against α-glucosidase and α-amylase for the management of post-prandial hyperglycemia via molecular modelling approaches. *Molecules* 27, 3888.

Pengsuparp, T., Serit, M., Hughes, S.H., Soejarto, D.D., and Pezzuto, J.M. (1996). Specific inhibition of human immunodeficiency virus type 1 reverse transcriptase mediated by soulattrolide, a coumarin isolated from the latex of Calophyllum teysmannii. *Journal of Natural Products* 59, 839-842.

Potdar, M.K., Mohile, S.S., and Salunkhe, M.M. (2001). Coumarin syntheses via Pechmann condensation in Lewis acidic chloroaluminate ionic liquid. *Tetrahedron Letters* 42, 9285-9287.

Pushpalatha, G., Pramod, N., Mahaboob Basha, G., Deepa, M., Neelaphar, P., and Jayakumar Swamy, B.H.M. (2016). Design, synthesis and anti-malarial activity of coumarin fused quinoline derivatives. *Journal of Pharmacy Research* 10, 437-441.

Rajtar, B., Skalicka-Woźniak, K., Polz-Dacewicz, M., and Głowniak, K. (2012). The influence of extracts from Peucedanum salinum on the replication of adenovirus type 5. *Archives of Medical Science* 8, 43-47.

Rani, A.J., Thomas, A., Arshad, M., Joseph, A., and Kuruvilla, M. (2022). The co-adsorption of thymohydroquinone dimethyl ether (THQ) and coumarin present in the aqueous extract of ayapana triplinervis on mild steel and its protection in hydrochloric acid up to 323 K: Computational and physicochemical studies. *RSC Advances* 12, 14328-14341.

Ren, Q.-C., Gao, C., Xu, Z., Feng, L.-S., Liu, M.-L., Wu, X., and Zhao, F. (2018). Bis-coumarin derivatives and their biological activities. *Current Topics in Medicinal Chemistry* 18, 101-113.

Riveiro, M.E., De Kimpe, N., Moglioni, A., Vazquez, R., Monczor, F., Shayo, C., and Davio, C. (2010). Coumarins: Old compounds with novel promising therapeutic perspectives. *Current Medicinal Chemistry* 17, 1325-1338.

Rodriguez-Dominguez, J.C., and Kirsch, G. (2006). Zirconyl chloride: A useful catalyst in the Pechmann coumarin synthesis. *Synthesis*, 1895-1897.

Rostom, B., Karaky, R., Kassab, I., and Sylla-Iyarreta Veitia, M. (2022). Coumarins derivatives and inflammation: Review of their effects on the inflammatory signaling pathways. *European Journal of Pharmacology* 922, 174867.

Russell, G.M., Masai, H., Terao, J., and Masai, H. (2022). Insulation of a coumarin derivative with rotaxane to control solvation-induced effects in excited-state dynamics for enhanced luminescence. *Physical Chemistry Chemical Physics* 24, 15195-15200.

Sandhu, S., Bansal, Y., Silakari, O., and Bansal, G. (2014). Coumarin hybrids as novel therapeutic agents. *Bioorganic & Medicinal Chemistry* 22, 3806-3814.

Sarker, S.D., and Nahar, L. (2017). Progress in the chemistry of naturally occurring coumarins. *Progress in the Chemistry of Organic Natural Products* 106, 241-304.

Sartiva, H., Ishida, M., Yoneyama, K., Nishiwaki, H., and Yamauchi, S. (2022). Plant growth suppressive activity of (R)-3-(7'-Aryl-9'-hydroxyprop-8'-yl)coumarin, structural isomer of Z-2-hydroxybenzylidene-γ-butyrolactone-type lignan. *Journal of Agricultural and Food Chemistry*.

Shen, W., Mao, J., Sun, J., Sun, M., and Zhang, C. (2013). Synthesis and biological evaluation of resveratrol-coumarin hybrid compounds as potential antitumor agents. *Medicinal Chemistry Research* 22, 1630-1640.

Singh, P., Ngcoya, N., Mopuri, R., Kerru, N., Manhas, N., Ebenezer, O., and Islam, M.S. (2018). α-Glucosidase inhibition, antioxidant and docking studies of hydroxycoumarins and their mono and bis O-alkylated/acetylated analogs. *Letters in Drug Design & Discovery* 15, 127-135.

Singh, R.K., Lange, T.S., Kim, K.K., and Brard, L. (2011). A coumarin derivative (RKS262) inhibits cell-cycle progression, causes pro-apoptotic signaling and cytotoxicity in ovarian cancer cells. *Investigational New Drugs* 29, 63-72.

Song, C., Li, X., Jia, B., Liu, L., Wei, P., Wang, F., Li, B.Y., Chen, C., Han, B., Song, C., et al. (2022). Comparative transcriptomics unveil the crucial genes involved in coumarin biosynthesis in peucedanum praeruptorum dunn. *Frontiers in Plant Science* 13, 899819.

Sui, Z., Luo, J., Yao, R., Huang, C., Zhao, Y., and Kong, L. (2019). Functional characterization and correlation analysis of phenylalanine ammonia-lyase (PAL) in coumarin biosynthesis from Peucedanum praeruptorum Dunn. *Phytochemistry* 158, 35-45.

Sun, Q., Li, X., Guo, Y., Qiu, Y., Luo, X., Liu, G., and Han, Y. (2022). Coumarin-based turn-on fluorescence probe with a large Stokes shift for detection of endogenous neutrophil elastase in live cells and zebrafish. *Spectrochimica Acta Part A: Molecular and Biomolecular Spectroscopy* 281, 121563.

Supuran, C.T. (2020). Coumarin carbonic anhydrase inhibitors from natural sources. *Journal of Enzyme Inhibition and Medicinal Chemistry* 35, 1462-1470.

Taştemel, A., Karaca, B.Y., Durmuş, M., and Bulut, M. (2015). Photophysical and photochemical properties of novel metallophthalocyanines bearing 7-oxy-3-(m-methoxyphenyl) coumarin groups. *Journal of Luminescence* 168, 163-171.

Thakur, A., Singla, R., and Jaitak, V. (2015). Coumarins as anticancer agents: A review on synthetic strategies, mechanism of action and SAR studies. *European Journal of Medicinal Chemistry* 101, 476-495.

Tong, X., Song, X., Wu, S., Zhang, N., Hao, L., Li, Z., Chen, S., and Hou, P. (2022). Construction of novel coumarin-carbazole-based fluorescent probe for tracking of endogenous and exogenous H_2S in vivo with yellow-emission and large Stokes shift. *Spectrochimica Acta Part A: Molecular and Biomolecular Spectroscopy* 279, 121445.

Utreja, D., Jain, N., and Sharma, S. (2018). Advances in synthesis and potentially bioactive of coumarin derivatives. *Current Organic Chemistry* 22, 2509-2536.

Valizadeh, H., and Vaghefi, S. (2009). One-pot Wittig and Knoevenagel reactions in ionic liquid as convenient methods for the synthesis of coumarin derivatives. *Synthetic Communications* 39, 1666-1678.

Verdia, P., Santamarta, F., and Tojo, E. (2017). Synthesis of (3-methoxycarbonyl) coumarin in an ionic liquid: An advanced undergraduate project for green chemistry. *Journal of Chemical Education* 94, 505-509.

Villa-Martinez, C.A., Ramos-Organillo, A.A., Martinez-Martinez, F.J., Magana-Vergara, N.E., Rodriguez, M., Mojica-Sanchez, J.P., Barroso-Flores, J., Barroso-Flores, J., and Padilla-Martinez, I.I. (2022). Synthesis, optical characterization in solution and solid-state, and DFT calculations of 3-acetyl and 3-(1'-(2'-phenylhydrazono)ethyl)-coumarin-(7)- substituted derivatives. *Molecules* 27, 3677.

Wang, H., Yao, M., and Xu, W. (2017). Syntheses and radiosensitization effect of mitochondria-targeted 6-(4-(dimethylamine) pyridine) methyl coumarin. *Journal of Radiation Research and Radiation Processing* 35, 3-8.

Wang, J., Lu, M.L., Dai, H.L., Zhang, S.P., Wang, H.X., and Wei, N. (2015). Esculetin, a coumarin derivative, exerts in vitro and in vivo antiproliferative activity against hepatocellular carcinoma by initiating a mitochondrial-dependent apoptosis pathway. *Brazilian Journal of Medical and Biological Research* 48, 245-253.

Wang, X., Gou, Z., Zuo, Y., and Lv, J.-J. (2022). A novel coumarin-TPA based fluorescent probe for turn-on hypochlorite detection and lipid-droplet-polarity bioimaging in cancer cells. *Spectrochimica Acta Part A: Molecular and Biomolecular Spectroscopy* 279, 121481.

Wang, Z., Kim, J.R., Wang, M., Shu, S., and Ahn, Y.J. (2012). Larvicidal activity of cnidium monnieri fruit coumarins and structurally related compounds against insecticide-susceptible and insecticide-resistant culex pipiens pallens and *Aedes aegypti*. *Pest Management Science* 68, 1041-1047.

Williams, K.J., and Gieling, R.G. (2019). Preclinical evaluation of ureidosulfamate carbonic anhydrase IX/XII inhibitors in the treatment of cancers. *International Journal of Molecular Sciences* 20, 6080.

Xi, G.-L., and Liu, Z.-Q. (2015). Coumarin-fused coumarin: Antioxidant story from N,N-dimethylamino and hydroxyl groups. *Journal of Agricultural and Food Chemistry* 63, 3516-3523.

Xia, Y.-L., Ge, G.-B., Wang, P., Liang, S.-C., He, Y.-Q., Ning, J., Qian, X.-K., Li, Y., and Yang, L. (2015). Structural modifications at the C-4 position strongly affect the glucuronidation of 6,7-dihydroxycoumarins. *Drug Metabolism and Disposition* 43, 553-560.

Xie, L., Chen, Y., Wu, W., Guo, H., Zhao, J., and Yu, X. (2012). Fluorescent coumarin derivatives with large stokes shift, dual emission and solid state luminescent properties: An experimental and theoretical study. *Dyes and Pigments* 92, 1361-1369.

Xu, Q., Yu, G., Liu, M., Peng, C., Banks, M.K., Xu, W., Wu, R., and Lu, Y. (2018). Coumarin-surfactant modified polyoxometalate catalyzed cross dehydrogenative coupling of benzyl alcohol with the para-C-H of unprotected aniline. *Catalysis Science & Technology* 8, 5133-5136.

Xu, R., Liu, Z., Liu, X., Wu, Y., and Liu, J. (2022). Lysosomal targeted cyclometallic iridium salicylaldehyde-coumarin schiff base complexes and anticancer application. *Frontiers in Chemistry* 10, 906954.

Yang, G., Jin, Q., Xu, C., Fan, S., Wang, C., and Xie, P. (2018). Synthesis, characterization and antifungal activity of coumarin-functionalized chitosan derivatives. *International Journal of Biological Macromolecules* 106, 179-184.

Yang, Q., Klan, P., Yang, Q., Klan, P., and Vana, J. (2022a). The complex photo-chemistry of coumarin-3-carboxylic acid in acetonitrile and methanol. *Photochemical & Photobiological Sciences* Doi:10.1007/s43630-022-00238-8.

Yang, Y., Tan, W., Zhang, J., Guo, Z., Jiang, A., and Li, Q. (2022b). Novel coumarin-functionalized inulin derivatives: Chemical modification and antioxidant activity assessment. *Carbohydrate Research* 518, 108597.

Yu, D., Suzuki, M., Xie, L., Morris-Natschke, S.L., and Lee, K.H. (2003). Recent progress in the development of coumarin derivatives as potent anti-HIV agents. *Medicinal Research Reviews* 23, 322-345.

Yu, Z., Guo, S., Wang, W., Qu, F., Chen, Y., Li, M., Ma, Y., and Liu, H. (2022a). Novel nitric oxide donor dinitroazetidine-coumarin hybrids as potent anti-intrahepatic cholangiocarcinoma agents. *Molecules* 27, 4021.

Yu, Z., Wang, K., Guo, S., Wang, W., Chen, Y., Li, M., Gu, Y., Ma, Y., Liu, H., and Wang, K. (2022b). Novel hybrids of 3-substituted coumarin and phenylsulfonylfuroxan as potent antitumor agents with collateral sensitivity against MCF-7/ADR. *Journal of Medicinal Chemistry* 65, 9328-9349.

Zhan, P., Chen, X., Li, D., Fang, Z., De Clercq, E., and Liu, X. (2013). HIV-1 NNRTIs: Structural diversity, pharmacophore similarity, and implications for drug design. *Medicinal Research RReviews* 33, E1-E72.

Zhang, S., Liu, W., Li, Z., Sun, Y., Feng, W., Zhang, M., and Yu, M. (2022a). Coumarin-based fluorescent probes toward viscosity in mitochondrion/lysosome. *Analytical Biochemistry* 652, 114752.

Zhang, Z., Zhang, Y., Xu, F., Li, K., Chang, L., Li, Y., Jiang, S., Gao, H., Kukic, P., Carmichael, P.L., et al. (2022b). Quantitative phosphoproteomics reveal cellular responses from caffeine, coumarin and quercetin in treated HepG2 cells. *Toxicology and Applied Pharmacology* 449, 116110.

Zhao, H., and Blagg, B.S. (2013). Inhibitors of the Hsp90 C-terminus. *Inhibitors of Molecular Chaperones as Therapeutic Agents* 37, 259-301.

Zhao, L., Zhang, S., Shan, C., Shi, Y., Wu, H., Wu, J., and Peng, D. (2021). De novo transcriptome assembly of angelica dahurica and characterization of coumarin biosynthesis pathway genes. *Gene* 791, 145713.

Zhou, R., Kusaka, E., Wang, Y., Zhang, J., Carrico-Moniz, D., and Webb, A. (2022a). Isoprenylated coumarin exhibits anti-proliferative effects in pancreatic cancer cells under nutrient starvation by inhibiting autophagy. *Anticancer Research* 42, 2835-2845.

Zhou, X., Gong, J., Zhuang, Y., and Zhu, F. (2022b). Coumarin protects cherax quadricarinatus (red claw crayfish) against white spot syndrome virus infection. *Fish & Shellfish Immunology* 127, 74-81.

Zhu, J.J., and Jiang, J.G. (2018). Pharmacological and nutritional effects of natural coumarins and their structure–activity relationships. *Molecular Nutrition & Food Research* 62, 1701073.

Chapter 4

Insights into the Structure-Activity Relationship of Alkynyl-Coumarinyl Ethers as Selective MAO-B Inhibitors Using Molecular Docking

Yassir Boulaamane[1,*]
Mohammed Reda Britel[1]
and Amal Maurady[1,2]

[1]Laboratory of Innovative Technologies, National School of Applied Sciences of Tangier, Abdelmalek Essaddi University, Tetouan, Morocco
[2]Faculty of Sciences and Techniques of Tangier, Abdelmalek Essaddi University, Tetouan, Morocco

Abstract

Coumarins are considered a highly privileged and versatile scaffold by medicinal chemists. A considerable number of studies have highlighted the synthesis and the various pharmacological activities of coumarins as promising drug candidates for treating neurodegenerative diseases such as Parkinson's and Alzheimer's disease. A wide range of compounds based on the coumarin ring system have been found to possess biological activities such as anticonvulsant, antiviral, anti-inflammatory, antibacterial, antioxidant as well as monoamine oxidase inhibitory properties. Their promise as a novel drug for neurodegenerative diseases is demonstrated by many drug candidates that made it to clinical trials such as nodakenin that have been potent for demoting memory impairment. This study focuses on some synthesized alkynyl-coumarinyl

---

[*] Corresponding Author's Email: boulaamane.yassir@etu.uae.ac.ma

In: The Chemistry of Coumarin
Editor: Scott R. Sheley
ISBN: 979-8-88697-560-4
© 2023 Nova Science Publishers, Inc.

ethers with promising MAO-B inhibitory activity and selectivity and aims to elucidates the molecular interactions of ether-connected coumarins behind obtaining remarkably high MAO-B selectivity using molecular docking. Structure-activity relationship analysis revealed a common interaction between the selective coumarin inhibitors consisting of hydrogen bonding with Tyr-188 and Cys-172. Our findings might open new opportunities to explore for developing novel highly selective MAO-B inhibitors for the treatment of neurodegenerative diseases.

Keywords: coumarin, neurodegenerative diseases, molecular docking, monoamine oxidase, structure-activity relationship

Introduction

Parkinson's disease (PD) is considered the second most frequent neurodegenerative disorder after Alzheimer's disease [1]. PD is defined by the progressive loss of dopaminergic neurons in the substantia nigra pars compacta (SNpc) of the mid brain [2]. Current treatments for PD are levodopa, which remains the gold standard, dopamine agonists and catechol-O-methyl transferase (COMT)/monoamine oxidase (MAO) inhibitors [3]. Monoamine Oxidase (MAO) (EC 1.4.3.4) is a mitochondrial flavoprotein attached to neurons outer-membrane that catalyses the oxidative deamination of neurotransmitters and biogenic amines [4]. MAO exists in two forms; MAO-A and MAO-B that share about 70% of their sequence identity, but differ in their tissue distribution, substrate, and inhibitor preferences [5]. MAO-A inhibitors are used as antidepressants, while selective MAO-B inhibitors have proven to be efficient in treating AD and PD symptoms. Moreover, they may act as neuroprotective agents by limiting the release of free radical species and hence decrease the progression of the disease [6,7].

MAO-A preferentially metabolizes serotonin while MAO-B preferentially deaminates 2-phenylethylamine and benzylamine. Dopamine, norepinephrine, and epinephrine are metabolized by both isoforms [8].

During aging, the expression of MAO-B increases in the brain and relates to an enhanced dopamine metabolism that produce reactive oxygen species (ROS) such as hydrogen peroxide (H_2O_2) resulting in oxidative damage and apoptotic signalling events [9].

MAO-A (PDB ID: 2Z5Y) is expressed as a monomer while MAO-B is formed of two monomers, both formed of a globular domain attached to the outer neuronal membrane through a C-terminal helix [10, 11]. The active site is located in the substrate fixing domain and is formed by the residues: Tyr-60, Pro-102, Pro-104, Leu-164, Phe-168, Leu-171, Cys-172, Ile-198, Ile-199, Gln-206, Ile-316, Tyr-326, Phe-343, Tyr-398 and Tyr-435 (10). Specific residues in MAO-B that are not present in MAO-A are: Leu-171, Cys-172, Ile-199 and Tyr-326 [11].

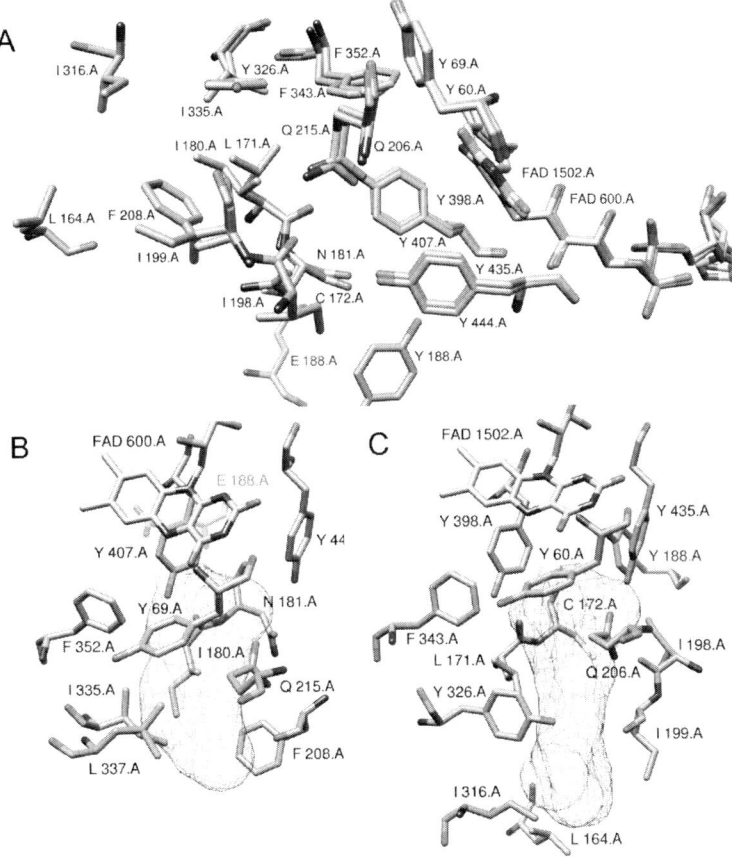

Figure 1. A) Superimposition of the binding sites of MAO-A (PDB ID: 2Z5Y) and MAO-B (PDB ID: 2V61); B) Binding surfaces of MAO-A. C) Binding surface of MAO-B shown in mesh representation.

Coumarins are considered a privileged scaffold in medicinal chemistry due to its peculiar physicochemical properties and the synthetic accessibility to transform it into a wide plethora of functionalized coumarins [12]. Coumarins have been broadly studied for developing new MAO inhibitors displaying a wide range of selectivity for MAO-B [13]. A recent study has reported that alkynyl coumarinyl ethers are able to inhibit MAO enzymes at nanomolar concentrations ranging from 0.58 nM to 1790 nM with a MAO-B selectivity reaching a value of over 3400-fold [14].

To develop new potent and highly selective MAO-B inhibitors, molecular modelling was used to get an insight on the possible molecular mechanisms of previously synthetised and biologically evaludated alkynyl-coumarinyl ethers [14].

Molecular docking study was carried out to investigate the structural conformations of the selected compounds with human MAO-B. Structure-activity relationship (SAR) analysis was conducted to identify the key molecular interactions that may enhance the selectivity for MAO-B.

Materials and Methods

Protein Preparation

The crystallographic structure of the human MAO-B (in complex with the selective inhibitor 7-(3-chlorobenzyloxy)-4-(methylamino) methyl-coumarin, C18) was obtained from the RCSB Protein Data Bank (http://www.rcsb.org/pdb/) (PDB ID: 2V61, resolution = 1.7 Å) and was prepared for molecular docking [11]. Co-crystallized ligand and water molecules were removed as they weren't involved in ligand binding. Chain B was removed and only chain A was kept along with the FAD cofactor as it plays an important role in catalysing the oxidative deamination of monoamines [15].

Chemical Structures Preparation

The selected coumarin derivatives were converted to chemical structures from their IUPAC nomenclature using MarvinSketch 20.9, 2020 (http://www.chemaxon.com) program. Explicit hydrogens and 3D coordinates were also generated. The Amber's antechamber module included with UCSF Chimera was used for energy minimisation of selected ligands, 100 steps of steepest descent minimization was performed, followed by 10 steps of conjugate gradient minimisation based on the AMBER ff14SB force field [16]. The chemical structures of coumarins are reported in Table 1 and 2.

Table 1. Chemical structures of R^1 derivatives and their MAO inhibitory activities [14]

Compound	Nomenclature	R^1	Structure	SI
1a	Methyl 2-oxo-7-(prop-2-ynyloxy)-2H-chromene-3-carboxylate	$CH_2C{\equiv}CH$		0.88
2a	Methyl 2-oxo-7-(but-3-ynyloxy)-2H-chromene-3-carboxylate	$(CH_2)_2C{\equiv}CH$		3.17
3a	Methyl 2-oxo-7-(pent-4-ynyloxy)-2H-chromene-3-carboxylate	$(CH_2)_3C{\equiv}CH$		11.56
4a	Methyl 2-oxo-7-(hex-5-ynyloxy)-2H-chromene-3-carboxylate	$(CH_2)_4C{\equiv}CH$		6.83
5a	Methyl 6-(hex-5-ynyloxy)-2-oxo-2H-chromene-3-carboxylate	$(CH_2)_4C{\equiv}CH$		81.30
6a	Methyl 7-(4-chlorophenethoxy)-2-oxo-2H-chromene3-carboxylate	$(CH_2)_2C_6H_4$-4-Cl		53
7a	Methyl 6-(4-chlorophenethoxy)-2-oxo-2H-chromene3-carboxylate	$(CH_2)_2C_6H_4$-4-Cl		>83.33
8a	Methyl 7-(3-fluorobenzyloxy)-2-oxo-2H-chromene-3carboxylate	$CH_2C_6H_4$-3-F		131.29
9a	Methyl 6-(3-fluorobenzyloxy)-2-oxo-2H-chromene-3carboxylate	$CH_2C_6H_4$-3-F		n/a

SI: Selectivity index (IC50 MAO-A/IC50 MAO-B).

Table 2. Chemical structures of R2 derivatives
and their MAO inhibitory activities [14]

Compound	Nomenclature	R2	Structure	SI
1b	3-(4-Methoxybenzoyl)-7-(hex-5-ynyloxy)-2H-chromen-2-one	COC6H4-4-OMe		>150
2b	3-(4-Methoxybenzoyl)-6-(hex-5-ynyloxy)-2H-chromen-2-one	COC6H4-4-OMe		—
3b	N-(2-Oxo-7-(hex-5-ynyloxy)-2H-chromen-3-yl)acetamide	NHCOMe		1.6
4b	N-(2-Oxo-6-(hex-5-ynyloxy)-2H-chromen-3-yl)acetamide	NHCOMe		>404.85
5b	7-(Hex-5-ynyloxy)-2H-chromen-2-one	H		140
6b	7-(Hex-5-ynyloxy)-3-(4-methoxyphenyl)-2H-chromen2-one	C6H4-4-OMe		>3378.37

SI: Selectivity index (IC$_{50}$ MAO-A/IC$_{50}$ MAO-B).

The molecules were regrouped into two groups: the first group (1a to 9a) contain R^1 derivatives with a methyl acetate moiety at C3 while the variation occurs at C6 or C7. Meanwhile, the second group (1b to 6b) contain R^2 derivatives with a hex-5-ynyloxy chain at C6 or C7 while the variation occurs at C3 as seen in Figure 2.

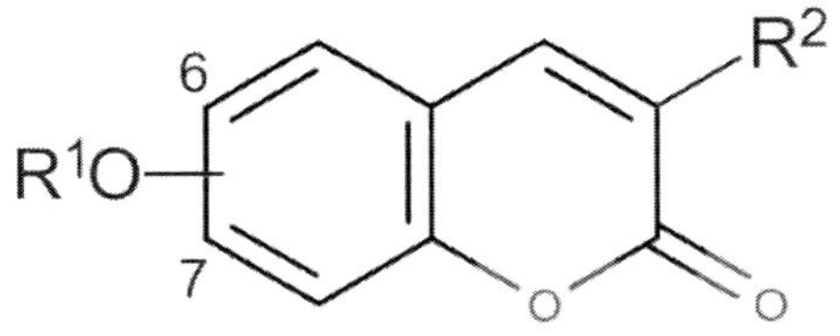

Figure 2. Coumarin scaffold with R^2 = CO_2Me at C3 (1a – 8a) and R^1 = $(CH_2)_4{\equiv}CH$ at C6 or C7 (1b – 6b).

Molecular Docking

Molecular docking was used for analysis of the interactions between the coumarin derivatives, and the active site of MAO-B. Docking simulation was performed by employing the AutoDock Vina 1.1.2 program [17]. The grid box was placed near the FAD with a spacing of 1 Å. Grid dimensions were chosen large enough (24 x 24 x 24 Å in x, y and z directions, respectively) to fit both cavities of the active site in the protein. The grid box was positioned in a way to cover the entire binding site and to allow larger molecules to dock properly (53 x 155 x 27 Å in x, y and z directions, respectively). Conformations of docked ligands were chosen according to their binding affinity and their conformation similarity to the native ligand.

Results

Molecular Docking Results

Conformations of docked compounds were ranked by their energies and then selected based on their similarity to the co-crystallized ligand which is also a coumarin derivative by mean of superposition. Hydrogen bonds were visualized using UCSF Chimera, Discovery Studio Visualizer was used to determine the nearby interacting hydrophobic amino acids [18]. Molecular docking results are shown in Table 3.

Table 3. Molecular docking results of selected ligands with MAO-B

Compound	ΔGb (kcal/mol)	Hydrogen bonds Residues	Bond length (Å)	Hydrophobic interactions
C18	-9.7	Tyr-435	3.0	Trp-119, Leu-164, Leu-167, Phe-168, Leu-171, Ile-199, Tyr-326, Phe-343
1a	-8.9	FAD-1502	2.3	Leu-171, Ile-316, Tyr-326
2a	-9.1	FAD-1502	2.4	Phe-168, Leu-171, Ile-199, Tyr-326, Tyr-398, Tyr-435
3a	-9.1	—	—	Leu-164, Leu-167, Phe-168, Leu-171, Ile-199, Ile-316, Tyr-398, Tyr-435
4a	-9.3	FAD-1502	2.5	Trp-119, Leu-171, Ile-199, Tyr-326, Tyr-398, Tyr-435
5a	-8.1	Cys-172	3.4	Phe-168, Leu-171, Ile-199, Gln-206, Tyr-398, Tyr-435
6a	-10.1	FAD-1502	2.3	Trp-119, Leu-164, Leu-171, Ile-199, Ile-316, Tyr-326
7a	-9.7	—	—	Trp-119, Leu-171, Ile-199, Ile-316, Tyr-326, Tyr-398, Tyr-435
8a	-10.8	FAD-1502	2.4	Leu-171, Ile-199, Ile-316, Tyr-326
9a	-9.7	FAD-1502 Tyr-188	2.6	Ile-199, Ile-316, Pro-102, Pro-104, Gly-434, Leu-171, Cys-172
1b	-8.0	—	—	Phe-103, Trp-119, Leu-164, Leu-167, Leu-171, Ile-199, Ile-316, Tyr-326, Tyr-398, Thr-399, Tyr-435
2b	-6.1	Cys-172	3.4	Trp-119, Leu-164, Phe-168, Leu-171, Ile-199, Ile-316, Tyr-326, Phe-343, Tyr-435
3b	-8.5	—	—	Phe-103, Trp-119, Leu-164, Leu-167, Leu-171, Ile-199, Ile-316, Tyr-398, Tyr-435
4b	-8.9	FAD-1502 Tyr-188	2.2 2.4	Trp-119, Leu-171, Ile-199, Tyr-326, Tyr-398, Tyr-435
5b	-9.2	—	—	Trp-119, Leu-171, Ile-199, Tyr-326
6b	-8.9	—	—	Pro-104, Trp-119, Leu-164, Leu-167, Phe-168, Ile-199, Ile-316, Leu-171, Ile-198, Cys-172, Tyr-326, Gln-206, Tyr-398, Tyr-435, Gly-434, FAD-1502

A good correlation ($R^2 = 0.535$) was established between docking results and Log of experimental IC_{50} values, which confirms the reliability of the molecular docking approach to study the mode of interaction of coumarin derivatives with MAO-B, the correlation plot and the equation used are shown in Figure 3.

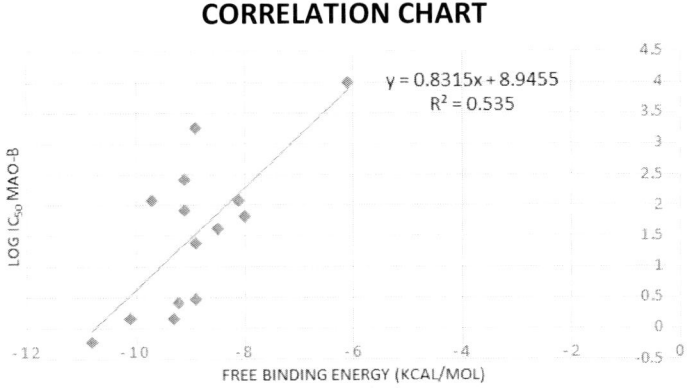

Figure 3. Correlation between docking free binding energy and experimental IC_{50} values.

The docking poses were visualized using UCSF Chimera visualization software and are shown in Figure 4.

An additional docking of safinamide (SAG) was conducted for comparison purpose and the result show that it binds to the Gln-206 as mentioned in the literature with a free binding energy of -10.1 kcal/mol [11]. Whereas the redocking of the co-crystallized coumarin derivative show that it forms a hydrogen bond with Tyr-435 of the aromatic cage. This difference may be due to the absence of water molecules during the docking process.

The fourteen compounds were separated into two groups: R^1 derivatives (1a-9a) with variation occurring in either C6 or C7 and R^2 derivatives (1b-6b) with variation occurring in C3.

The docking poses were visualized using UCSF Chimera visualization software and are shown in Figure 4.

An additional docking of safinamide (SAG) was conducted for comparison purpose and the result show that it binds to the Gln-206 as mentioned in the literature with a free binding energy of -10.1 kcal/mol [11]. Whereas the redocking of the co-crystallized coumarin derivative show that it forms a hydrogen bond with Tyr-435 of the aromatic cage. This difference may be due to the absence of water molecules during the docking process.

The fourteen compounds were separated into two groups: R^1 derivatives (1a-9a) with variation occurring in either C6 or C7 and R^2 derivatives (1b-6b) with variation occurring in C3.

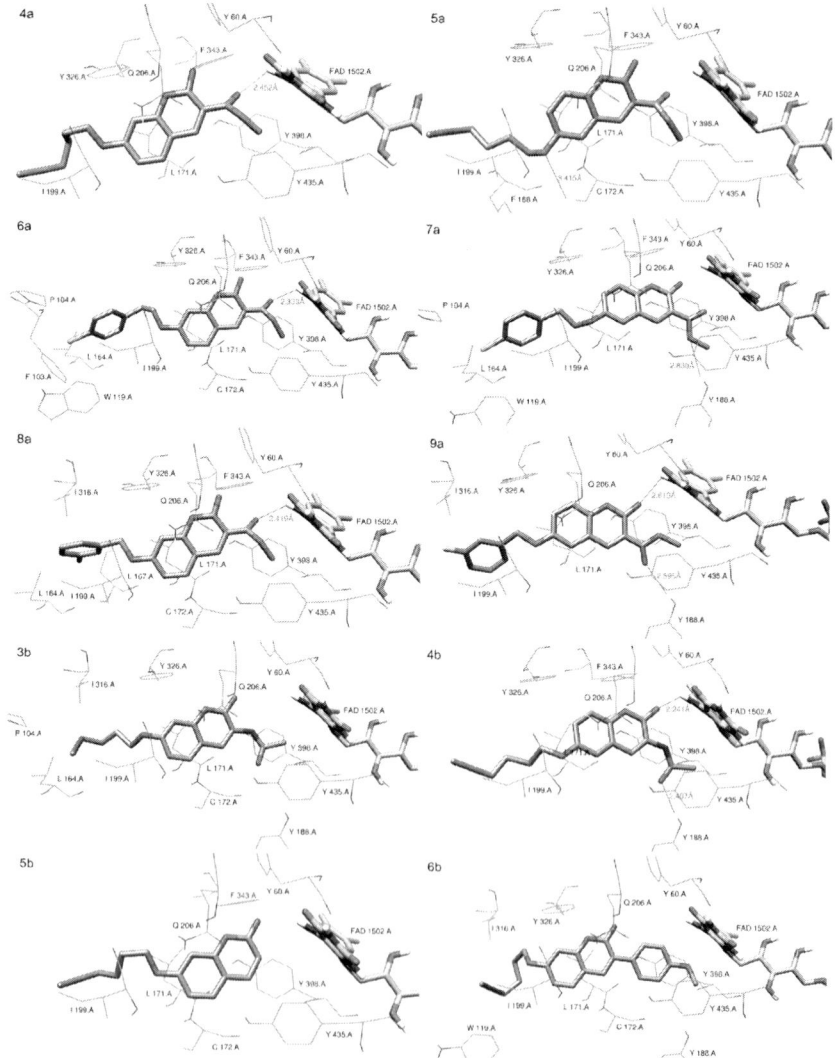

Figure 4. Docking poses of the selected coumarin derivatives within MAO-B active site, hydrogen bonds are shown in green lines.

Structure-Activity Relationship Analysis

To identify the structural requirements for coumarin derivatives to potently and selectively inhibit MAO-B, the first four molecules (1a-4a) were modified by adding a methyl group in the ether chain in each molecule. The ether chain elongation which resulted in an increased molecule length has been shown to be favourable for MAO-B inhibition. Molecular docking showed a significant increase in hydrophobic interactions. MAO-B active site is a long cavity and hence the elongated ether chain allowed the ligand to occupy both cavities and interact with the FAD cofactor through hydrogen bonding as demonstrated in the compound 4a. This compound is considered a dual inhibitor with IC_{50} values of 9.64 nM and 1.41 nM for MAO-A and MAO-B respectively [14].

The compound 5a which is a C7-isomer of 4a has been shown to be slightly less potent with an IC_{50} of 123 nM but more selective towards MAO-B (SI > 81). The structural analysis reveals that the ether chain in the C6-isomer is directed towards the bottom of the entrance cavity and forms a hydrogen bond with the residue Cys-172 which is not present in MAO-A suggesting that this residue may play a role in MAO-B selectivity.

We note that this increase in selectivity due to the replacement of the ether chain in C6 is also noted in the compound 7a. The docking result shows that it binds to the aromatic residue Tyr-188 which is located in the bottom of the aromatic cage through hydrogen bonding. We note that this residue is replaced with a glutamic acid in MAO-A and thus may be involved in MAO-B selectivity.

Lastly, the compound 8a was modified by adding a 4-fluorobenzyloxy moiety in C7, according to the docking result, this ligand is the most stable amongst the selected compounds displaying the lowest binding affinity (-10.8 kcal/mol). We note that this compound was also the most potent with an IC_{50} value of 0.58 nM according to the experimental study [14].

In the compound 9a, a C6-isomer of 8a was modelled and docked for comparison purpose. Our result show that it binds in a similar way to the compound 7a, forming two hydrogen bonds with FAD cofactor and Tyr-188. Detailed SAR analysis is illustrated in Figure 5.

In the second group, the C6-isomer 2b bearing the 1-(4-methoxyphenyl) ethan-1-one at C3 is reported for losing inhibitory activity for both MAO-A and MAO-B, the docking result also shows that this compound binds to MAO-B with the highest binding affinity among the selected compounds. However, the addition of the N-methyl acetamide moiety in the isomers 3b and 4b has been shown to correlate well with the previous suggestions. The C6-isomer 4b is 400 times more selective for MAO-B compared to its C7-isomer. The binding mode is similar to the previous C6-isomers. Two hydrogen bonds were visualized involving the FAD cofactor and the residue Tyr-188.

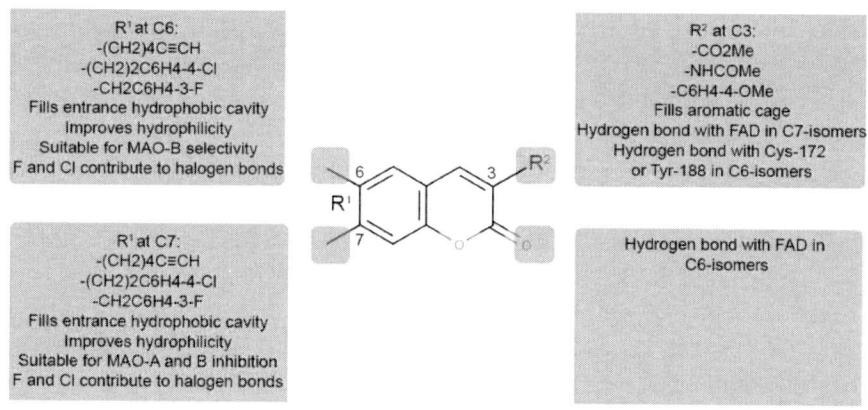

Figure 5. Structure-activity relationship (SAR) analysis of alkynyl coumarinyl ethers.

Finally, the last compounds bearing the hex-5-ynyloxy chain at C7 were compared by adding a 1-methoxy-4-methylbenzene moiety at C3 in the compound 6b while leaving it blank in the compound 5b. The docking result of the compound 6b shows that it binds to MAO-B with a free binding energy of -8.9 kcal/mol. Its experimental IC_{50} value is estimated to 2.96 nM and displays a MAO-B selectivity of over 3400 [14].

The structural analysis shows that the aromatic ring at C3 in the compound 6b occupies the aromatic cage and it's stabilized between the two residues Tyr-398 and Tyr-435. Meanwhile the coumarin ring is involved in π-stacking interactions with the gating residues Ile-199 and Tyr-326. Other hydrophobic interactions are shown in Table 3.

Discussion

Based on previously reported experimental data, it was confirmed that C7-isomers of coumarins tend to be more potent towards MAO-B, meanwhile the C6-isomers are slightly less potent but tend to be more selective towards MAO-B isoform [14]. We noticed that this hypothesis is applied to the compounds: 4a, 6a and 3b and their respective C6-isomers: 5a, 7a and 4b which displayed a MAO-B selectivity of approximately 80, 80 and 400-fold respectively. Due to the absence of compound 8a isomer, we used molecular modeling to design a C6-isomer of this compound and was docked within MAO-B active site. Structural analysis revealed that it binds in a similar way to other C6-isomers, with the 4-fluorobenzyloxy moiety directed towards the bottom of the cavity and engaging the residue Tyr-188 in a hydrogen bond with the oxygen of the methyl acetate moiety at C3, meanwhile the oxygen of coumarin scaffold established a hydrogen bond with the cofactor FAD as observed in most compounds. However, an *in vitro* inhibition assay is required to determine its MAO-B selectivity.

Structural analysis of the most selective compound, 6b, revealed that it does not bind to MAO-B active site through any hydrogen bonds but establishes hydrophobic interactions involving various residues of the hydrophobic pocket which seems to be more favorable for the stability of the protein-ligand complex than any other interactions such as hydrogen or halogen bonds. Furthermore, the long shape of the molecule plays a role in its selectivity as the differences between MAO-A and MAO-B are mainly related to the shape and the flexibility of their active site cavities. The long and narrow cavity of MAO-B makes it preferentially bind long inhibitors which forces a conformational change of the gating residue Ile-199 and fuses the two cavities into one [19-21]. The absence of this mechanism in MAO-A isoform further strengthen this hypothesis and could explain why such inhibitors tend to be more selective towards MAO-B.

Moreover, the molecular docking study further confirmed that all coumarin derivatives bind non-covalently to MAO-B active site and the triple bond of the ether chain doesn't bind to the cofactor FAD as such in irreversible inhibitors.

Conclusion

The current study aimed to shed a light on the mode of interaction of previously reported alkynyl coumarinyl ethers at the molecular level. It was found that C6-isomers are more selective towards MAO-B compared to their respective C7-isomers. Structure-activity relationship revealed that the loss of activity towards MAO-A of these compounds may be due to the bulky side chain of Phe-208 which is replaced by the gating residue Ile-199 that displays a conformational change depending on the nature of the inhibitor. Among the studied ligands, the compound 6b is considered the best drug-candidate among the fourteen compounds which needs more focus for the development of new antiparkinsonian drugs in respect to its drug likeness, high potency and selectivity for MAO-B.

Disclaimer

None.

References

[1] Noda, S., Sato, S., Fukuda, T., Tada, N., Uchiyama, Y., Tanaka, K., & Hattori, N. (2020). Loss of Parkin contributes to mitochondrial turnover and dopaminergic neuronal loss in aged mice. *Neurobiology of Disease,* 136, 104717

[2] Poewe, W., & Mahlknecht, P. (2020). Pharmacologic Treatment of Motor Symptoms Associated with Parkinson Disease. *Neurologic Clinics*, 38(2), 255-267.

[3] Dorsey, E. R., Elbaz, A., Nichols, E., Abd-Allah, F., Abdelalim, A. Adsuar, J. C., Mustafa Geleto Ansha, Carol Brayne, Jee-Young J Choi, Daniel Collado-Mateo, Nabila Dahodwala, Huyen Phuc Do, Dumessa Edessa, Matthias Endres, Seyed-Mohammad Fereshtehnejad, Kyle J Foreman, Fortune Gbetoho Gankpe, Rahul Gupta, Samer Hamidi, Graeme J. Hankey, Simon I. Hay, Mohamed I Hegazy, Desalegn T. Hibstu, Amir Kasaeian, Yousef Khader, Ibrahim Khalil, Young-Ho Khang, Yun Jin Kim, Yoshihiro Kokubo, Giancarlo Logroscino, João Massano, Norlinah Mohamed Ibrahim, Mohammed A. Mohammed, Alireza Mohammadi, Maziar Moradi-Lakeh, Mohsen Naghavi, Binh Thanh Nguyen, Yirga Legesse Nirayo, Felix Akpojene Ogbo, Mayowa Ojo Owolabi, David M. Pereira, Maarten J Postma, Mostafa Qorbani, Muhammad Aziz Rahman, Kedir T. Roba, Hosein Safari, Saeid Safiri, Maheswar Satpathy, Monika Sawhney, Azadeh Shafieesabet, Mekonnen Sisay Shiferaw, Mari Smith, Cassandra E I Szoeke, Rafael Tabarés-Seisdedos, Nu Thi Truong, Kingsley Nnanna Ukwaja, Narayanaswamy Venketasubramanian, Santos Villafaina, Kidu Gidey Weldegwergs, Ronny Westerman, Tissa Wijeratne, Andrea S. Winkler, Bach Tran Xuan, Naohiro Yonemoto, Valery L Feigin, Theo Vos, Christopher J L Murray. (2018). Global, reginal, and national burden of Parkinson's disease, 199-2016: a systematic analysis for the Global Burden of Disease Study 2016. *The Lancet Neurology*, 17(11), 939-953.

[4] Youdim, M. B., Edmondson, D., & Tipton, K. F. (2006). The therapeutic potential of monoamine oxidase inhibitors. *Nature reviews neuroscience*, 7(4), 295-309.

[5] Shih, J. C., Chen, K., & Ridd, M. J. (1999). Monoamine oxidase: from genes to behavior. *Annual review of neuroscience*, 22(1), 197-217.

[6] Youdim, M. B., Edmondson, D., & Tipton, K. F. (2006). The therapeutic potential of monoamine oxidase inhibitors. *Nature reviews neuroscience*, 7(4), 295-309.

[7] Culpepper, L. (2013). Reducing the burden of difficult-to-treat major depressive disorder: revisiting monoamine oxidase inhibitor therapy. *The primary care companion for CNS disorders*, 15(5).

[8] Youdim, M. B., Gross, A., & Finberg, J. P. (2001). Rasagiline [N-propargyl-1R (+)-aminoindan], a selective and potent inhibitor of mitochondrial monoamine oxidase B. *British journal of pharmacology*, 132(2), 500-506.

[9] Jenner, P., & Olanow, C. W. (1996). Oxidative stress and the pathogenesis of Parkinson's disease. *Neurology*, 47(6 Suppl 3), 161S-170S.

[10] Son, S. Y., Ma, J., Kondou, Y., Yoshimura, M., Yamashita, E., & Tsukihara, T. (2008). Structure of human monoamine oxidase A at 2.2-Å resolution: the control of opening the entry for substrates/inhibitors. *Proceedings of the National Academy of Sciences*, 105(15), 5739-5744.

[11] Binda, C., Wang, J., Pisani, L., Caccia, C., Carotti, A., Salvati, P., Dale E. Edmondson, & Mattevi, A. (2007). Structures of human monoamine oxidase B complexes with selective noncovalent inhibitors: safinamide and coumarin analogs. *Journal of medicinal chemistry*, 50(23), 5848-5852.

[12] Stefanachi, A., Leonetti, F., Pisani, L., Catto, M., & Carotti, A. (2018). Coumarin: A natural, privileged and versatile scaffold for bioactive compounds. *Molecules*, 23(2), 250.

[13] Gnerre, C., Catto, M., Leonetti, F., Weber, P., Carrupt, P. A., Altomare, C., A Carotti & Testa, B. (2000). Inhibition of monoamine oxidases by functionalized coumarin derivatives: biological activities, QSARs, and 3D-QSARs. *Journal of medicinal chemistry*, 43(25), 4747-4758.

[14] Mertens, M. D., Hinz, S., Müller, C. E., & Gütschow, M. (2014). Alkynyl-coumarinyl ethers as MAO-B inhibitors. *Bioorganic & medicinal chemistry*, 22(6), 1916-1928.

[15] Edmondson, D. E., Binda, C., & Mattevi, A. (2004). The FAD binding sites of human monoamine oxidases A and B. *Neurotoxicology*, 25(1-2), 63-72.

[16] Pettersen, E. F., Goddard, T. D., Huang, C. C., Couch, G. S., Greenblatt, D. M., Meng, E. C., & Ferrin, T. E. (2004). UCSF Chimera - a visualization system for exploratory research and analysis. *Journal of computational chemistry*, 25(13), 1605-1612.

[17] Trott, O., & Olson, A. J. (2010). AutoDock Vina: improving the speed and accuracy of docking with a new scoring function, efficient optimization, and multithreading. *Journal of computational chemistry*, 31(2), 455-461.

[18] Systèmes, D. (2016). *Biovia, discovery studio modeling environment*. Dassault Systèmes Biovia: San Diego, CA, USA.

[19] Finberg, J. P., & Rabey, J. M. (2016). Inhibitors of MAO-A and MAO-B in psychiatry and neurology. *Frontiers in pharmacology*, 7, 340.

[20] Boulaamane, Y., Ahmad, I., Patel, H., Das, N., Britel, M. R., & Maurady, A. (2022). Structural exploration of selected C6 and C7-substituted coumarin isomers as

selective MAO-B inhibitors. *Journal of Biomolecular Structure and Dynamics*, 1-15.

[21] Boulaamane, Y., Ibrahim, M. A., Britel, M. R., & Maurady, A. (2022). *In silico studies of natural product-like caffeine derivatives as potential MAO-B inhibitors/AA2AR antagonists for the treatment of Parkinson's disease*. *Journal of Integrative Bioinformatics*.

Chapter 5

Phytochemistry and Pharmacological Actions of Coumarin

Ramesh Vimalavathini [1,*], PhD
and Gnanakumar Prakash Yoganandam [2], PhD

[1]Department of Pharmacology, College of Pharmacy, Mother Theresa Post Graduate & Research Institute of Health Sciecnes, Pondicherry, India
[2]Department of Pharmacognosy, College of Pharmacy, Mother Theresa Post Graduate & Research Institute of Health Sciecnes, Pondicherry, India

Abstract

Coumarins ($2H$-1-benzopyran-2-one) are the leading group of benzopyran derivatives that originate in plants. Coumarin name, is derived from a French term, *Coumarou* an aromatic, colorless compound, and was first isolated from the Tonka bean (*Coumarouna odorata*, Wild, family Fabaceae) in 1820. Coumarins are conveyed in about 150 diverse species distributed over nearly 30 different families, of which a few important are Apiaceae, Asteraceae, Rutaceae, Umbelliferae and Clusiaceae. Their structure consists of two six-membered rings with lactone carbonyl groups. Most coumarin compounds are thermally stable and have distinguished optical activity. They are biosynthesized from the phenyl propanoid pathway via ortho-hydroxylation. After hydroxylation, trans/cis isomerization and lactonization occur, resulting in the production of their respective coumarins. They have been the centre of attraction for medicinal chemist for the past few decades owing to their profound pharmacological activities like anti-microbial, anti-oxidant, anti-inflammatory, analgesic, anti-cancer, anti-malarial, anti-

---

* Corresponding Author's Email: vimalavathini@gmail.com

In: The Chemistry of Coumarin
Editor: Scott R. Sheley
ISBN: 979-8-88697-560-4
© 2023 Nova Science Publishers, Inc.

hyperlipidemic, anti-epileptic, anti-parkinsonian, anti-hepatitis, anti-coagulant, enzyme inhibiton and vasorelaxant properties. This chapter is a write up on the occurrence, phytochemistry and therapeutic actions of coumarins.

1. Occurrence, Distribution, Chemistry, Classification and Biosynthetic Pathway of Coumarins

Occurrence

Coumarins (2H-1-benzopyran-2-one) (**1**) (Figure 1) are the largest group of 1-benzopyran derivatives found in plants. The initial member of this group of compound is coumarin which is a fragrant, colorless compound first isolated from the Tonka bean (*Coumarouna odorata*, Wild, family Fabaceae) in 1820. The name coumarin originates from a French word for the tonka bean, *coumarou*. Coumarins and its several derivatives, with umbelliferone (7-hydroxycoumarin) being the most common one, have been conveyed from various natural sources.

The families Apiaceae, Asteraceae, and Rutaceae are the three most important plant sources of coumarins. Coumarins are reported in about 150 different species distributed over nearly 30 different families, of which a few imperative ones are Rutaceae, Umbelliferae, Clusiaceae, Guttiferae, Caprifoliaceae, Oleaceae, Nyctaginaceae and Apiaceae. In general, they are classified into simple, simple prenylated, simple geranylated, furano, pyrano, sesquiterpenyl and oligomeric coumarins (Sarker and Nahar, 2017). More than 1300 coumarins have been acknowledged as secondary metabolites from plants, bacteria and fungi. Natural compound coumarins have been showing to considerable phytochemical and pharmacological consideration in the last few decades.

It is revealed that over the past three years, above 400 coumarins have been described in scientific periodicals (Martins Borges et al., 2009).

Distribution

It is also existing in high amount in Vanilla grass (*Anthoxanthum odoratum*) (Tava, 2001), Sweet clover (*Melilotus officinalis*) (Luo et al., 2016), Cassia cinnamon (*Cinnamomum cassia*) (Blahová and Svobodová, 2012), in the extracts of Water-willow (*Justicia pectoralis*) (Leal et al., 2000), and a large number of cherry blossom trees (genus Prunus or Prunus subg. Cerasus) (Poonam et al., 2011). The Apiaceae family includes several species that are high in coumarin content, including Pragos Lindl., Ferula L., Heracleum L., Pachypleurum Hoff., Conioselinum Fisch., Libanotis L., and Seseli L. (Sharopov and Setzer, 2018). Many plants contain different concentrations of coumarin. Natural coumarins are abundant in tonka beans, liquorice, and cassia cinnamon. Some Cherry blossoms, Strawberries and Apricots contains coumarin in lesser quantities. Regardless of its sweet odour, animals incline to avoid plants that contains coumarin due to their bitter taste (Menezes and Diederich, 2019).

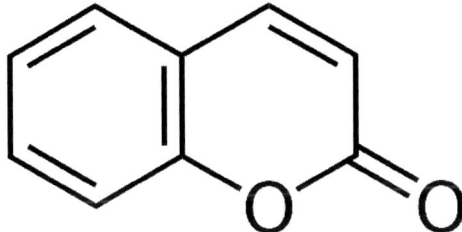

Figure 1. Chemical structure of Coumarin.

Chemistry

Coumarins belong to the family benzopyrone, commonly found in many medicinal plants (Figure 1). Their structure comprises of double six-membered rings with lactone carbonyl groups. The chemical compound coumarin, also known as 2H-chromen-2-one, has the formula $C_9H_6O_2$. Its molecule can be described as a benzene molecule with two adjacent hydrogen atoms replaced by a lactone-like chain − = -O-, forming a second six-membered heterocycle that shares two carbons with the benzene ring. Most coumarin compounds are heat stable and have distinguished optical activity (Kadhum et al., 2011). The coumarin biosynthesis takes place in encompassing multiple p450 enzymes (Bourgaud et al., 2006). A crucial step in the manufacture of natural coumarins in plants is ortho-hydroxylation. (Shimizu et al., 2014).

Classification

Natural coumarins are mainly classified into six types based on the chemical structure of the compounds as described in Table 1.

Table 1. Coumarins- classification based on chemical structures

S.No	Type of Coumarins	Chemical Structure	Examples
1.	Simple coumarins		Coumarin Esculetin Ammoresinol Umbelliferone
2.	Furano coumarins		Imperatorin Psoralen Bergapten marmelosin
3.	Dihydrofurano coumarins		Anthogenol Felamidin Marmesin
4.	Pyrano coumarins		Agasyllin Xanthyletin Inophyllum Calanolide
5.	Phenyl coumarins		Disparpropylinol Dispardiol
6.	Bicoumarins		Dicoumarol

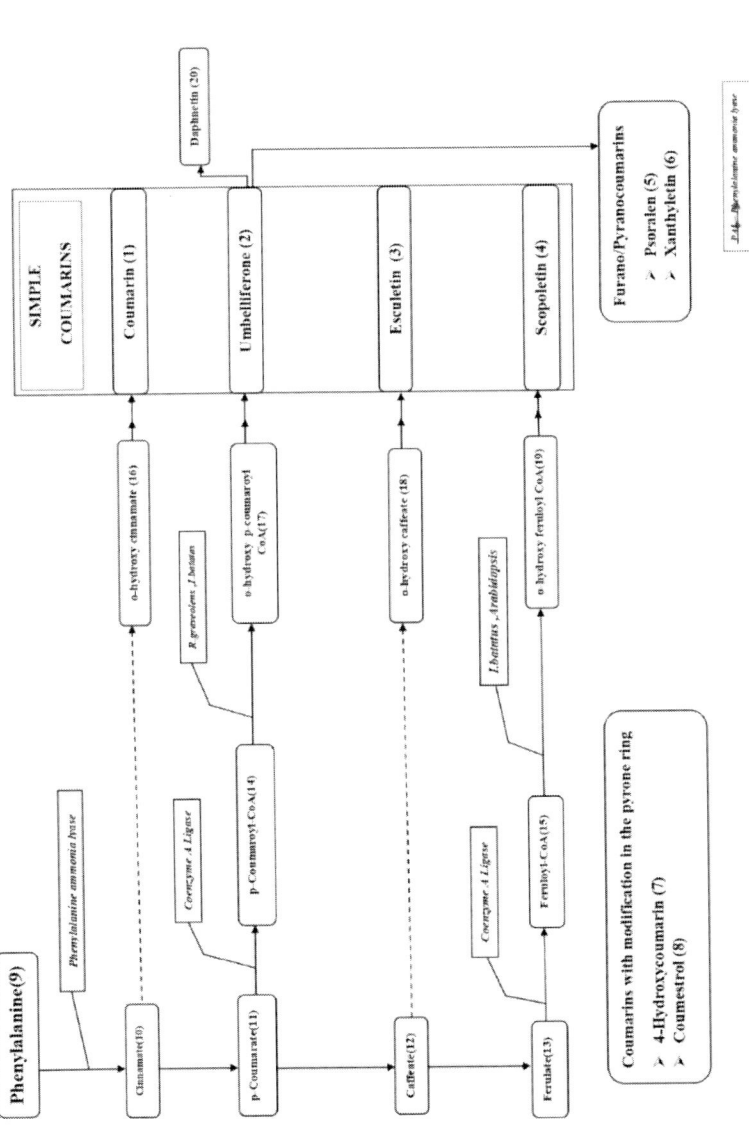

Figure 2. Biosynthetic pathway of Coumarin in plants.

Biosynthesis

Simple coumarins have modifications to their benzene rings, including coumarin (**1**), umbelliferone (**2**), esculetin (**3**), and scopoletin (**4**). Through the ortho-hydroxylation of cinnamate (**10**), p-coumarate (**11**), caffeate (**12**), and ferulate (**13**), they are each biosynthesized from the phenylpropanoid pathway. The ortho-positions are shown by arrows. Also noted are the oxygen atoms produced by ortho-hydroxylation. Functional analysis was done on the ortho-hydroxylases from Arabidopsis (AtF6′H1), Ruta graveolens (RgC2′H), and Ipomoea batatas (Ib1 and Ib2). While RgC2′H and Ib2 were able to respond to both feruloyl-CoA (**15**) and p-coumaroyl-CoA (**14**) as the substrates, AtF6′H1 and Ib1 catalyse ortho-hydroxylation of feruloyl-CoA (**15**).

After hydroxylation, trans/cis isomerization and lactonization occurs, following on in the construction of their respective coumarins. A crucial step in the biosynthesis of prenylcoumarins is the production of umbelliferone (**2**), from which furanocoumarins and pyranocoumarins (such as psoralen and xanthyletin, respectively) are generated.

No report has described cloning and functional analysis of the hydroxylases that bring together an ortho-hydroxy group to cinnamate and caffeate to form coumarin (**1**) and esculetin (**3**), respectively (hashed arrows). Different processes are thought to have produced coumarins with the pyrone ring replaced (Shimizu et al., 2014).

2. Pharmacological Actions of Coumarin

Coumarin has become an attractive moiety for medicinal research owing to its extensive distribution, stability, solubility, and propensity for easy chemical modification with diverse pharmacological effects such as antioxidant, anti-inflammatory, analgesic, anticancer, antidepressant, anti-Alzheimer, anti-epileptic, anti-coagulant, anti-diabetic, anti-hyperlipidemic, anti-asthmatic, vasorelaxant, antibacterial, antifungal, antimalarial, antiviral, antiretroviral, and anti-tuberculosis. They have wide applications in food, agriculture, perfumes, pharmaceuticals and cosmetics industries. Several members of this class are currently being clinically used as anticoagulants, (Warfarin, Phenprocoumon, difenacoum, dicumarol, acenocoumarol and tecarfarin) antibiotics, (Novobiocin and Armillarisin A) anti-dementia, (Ensaculin) antispasmodic, (Hymecromone) anti-cancer, (auraptene) coronary vasodilator, (carbochromen) antifungal, (scopoletin) and platelet aggregation inhibitor (cloricromen). The forthcoming paragraphs reviews various pharmacological profiles of natural coumarins and its derivatives (Keri RS et al., 2022, Sharanabasappa B et al., 2012).

Anti-Oxidant Action

Coumarins exhibit antioxidant activity by inhibition of reactive oxygen species (ROS) synthesis, producing antioxidant enzymes, direct free radical scavenging and chelating activity. The anti-inflammatory, neuroprotective and anticancer activity of coumarin may be associated with its antioxidant and free radical scavenging effect. Studies show that coumarins with catechol structure, coumarinyl thiazoles, coumarinyl thiadiazine, 4 methyl coumarin has promising antioxidant activity (Bubols GB et al., 2013).

Anti-Inflammatory Action

Inflammation is associated with multiple chronic diseases like cancer, diabetes mellitus, autoimmune, cardiovascular and neurodegenerative diseases. Anti-inflammatory ability of coumarin is owing to its anti-oxidant, and ability to hamper synthesis and secretion of plasma proteases, cyclooxygenases, caspases, nuclear factor-KB, kinases, prostaglandins leukotrienes, histamine, serotonin, nitric oxide, cytokines, chemokines and colony-stimulating factors. Literature review reflect that coumarin derivatives such as coumarin-3-carboxamides, 4-aryl coumarins, coumarinyl benzimidazole, 7-alkyloxy coumarins and acyl dihydroxy coumarin derivatives exhibit synergistic antioxidant aas well as anti-inflammatory activities. Natural coumarins namely, imperatorin, umbelliferone, visniadin, scopoletin, esculetin, columbiatnetin and marmin are potent anti-inflammatory agents (Bansal Y et al., 2013).

Anticancer Action

Disproportionate metabolic activity and compromised mitochondrial function of cancer cells succumbs itself to surplus free radical generation, inflammation, apoptosis, tumorigenesis, cell survival, proliferation, multidrug resistance (MDR) and metastasis. Natural coumarins namely grandivittin, agasyllin, fraxin, aegelinol, esculetin and osthole and synthetic derivatives such as coumarin-3-carboxamides, azocoumarin, bromo substiued coumarinyl hydrazides, coumarinyl pyrazoline, chalcone coumarins hybrids, 4 methyl coumarins, hydroxycoumarin displayed potent anticancer activity. Various studies has revealed that coumarins can impede the generation, propagation and metastasis of various cancer by multiple mechanism of action such as antioxidant, anti-inflammatory properties, inhibition of carbonic anhydrase, targeting various intracellular signaling pathways, modulating cell apoptosis, impeding multidrug resistance (MDR), mitotic spindle arrest and arresting tumor angiogenesis.

Carbonic anhydrases are unregulated in cancer propagating metastasis and coumarin analog, psoralidin can prevent Notch-1 mediated tumor growth and metastasis in carbonic anhydrases dependent cancers. Aberration of phosphatidylinositol 3-kinase (PI3K)/ protein kinase B (Akt)/mammalian target of rapamycin (mTOR) pathway is carcinogenic and imperatorin displayed anti-cancer effect by hindering PI3K, Akt and mTOR phosphorylation. Studies reflect that coumarins can also regulate B-cell lymphoma 2 (Bcl- 2) family of anti-apoptotic protein thereby preventing tumor proliferation in leukaemia, ovarian, prostrate, pancreas, cervical, breast and hepatocellular carcinoma.

MDR contributes to tumor recurrence and is mediated by p-glycoprotein (P-gp) and multidrug resistance associated protein 2 (MRP--2). Various studies reveals that coumarins can impede P-gp or MRP -2 in breast, cervical, colorectal and liver cancer cells. Esculetin inhibits cancer metastasis by deterring signal transducer and activator of transcription (STAT) phosphorylation and transport to the nucleus. Osthole impedes matrix metalloproteinases preventing cancer metastasis. Coumarin compounds such as aesculetin, umbelliferone, and scopoletin exhibited anticancer properties inhibiting NAD(P)H quinone oxidoreductase-1. Shikonin, a coumarin derivative can arrest glycolysis and stimulates apoptosis and cell adhesion in cancer cells by modulating hypoxia-inducible factor1-alpha degradation. In addition to anti-cancer properties studies show that coumarins can also mitigate radiotherapy induced sialadentis and mucositis (Wu Y et al., 2020).

Anti-Diabetic and Hypolipidemic Action

Dyslipidemia and diabetes go hand in hand and hence antidiabetic effects of coumarins are also due to its antioxidant, anti-inflammatory and anti-hyperlipidemic activity. Studies show that both natural coumarins such as umbelliferone, pyranocoumarins, furocoumarins bicoumarins and synthetic coumarin derivatives exerts anti diabetic and anti- hyperlipidemic effect by multimodal action. They stimulate beta cells, prevent insulin resistance, improves uptake and utilisation of lipids and sugars by cells, activates lipolytic enzymes, glucagon-like peptide-1 (GLP-1) modulation, trigger AMP-activated protein kinase (AMPK), and inhibits a-glucosidases along with anti-inflammatory antioxidant and anti-apoptosis action.

Osthole significantly activates peroxisome proliferator-activated receptor (PPAR) and impairs tyrosine phosphatase activity. Scopoletin activates Akt phosphorylation and PPAR expression along with impairment of advanced glycation endproducts production. Fraxetin, modulate the activity of enzymes of carbohydrate metabolism. Propensity of coumarins to intervene Toll-like receptor 4 (TLR4) signalling pathway and T-lymphocyte proliferation has made it an attractive scaffold for type 1 diabetes mellitus treatment. Aculeatin, promotes differentiation and lipolysis and is an efficient antidiabetic and anti-hyperlipidemic coumarin (Li H. et al., 2017].

Anticoagulant Action

Warfarin is the first generation coumarins used as an anticoagulant. 4-aminocoumarin, 4-hydroxy coumarins derivatives inhibit coagulation and platelet aggregation. Coumarins block multiple steps in the coagulation cascade and inhibits the production of vitamin K-dependent coagulation factors like II, VII, IX, and X in the liver and impedes ADP-induced platelet activation and aggregation (Srikrishna D et al., 2016, Lu PH et al., 2022). Studies show that furocoumarins exhibits cardio protective and antihypertensive properties (Sharifi-Rad J et al., 2021, Razavi BM et al., 2015) owing to their vasorelaxant activity (Campos-Toimil M. et al., 2002).

Neuroprotective Action

Neurodegenerative diseases are characterized by progressive dysfunction and neuronal loss. Coumarin seems to be effective in alleviating neurodegenerative diseases by antioxidant action, inhibition of monoamine oxidase-B (MAO-B), prevention of amyloid beta aggregation, and counteracting neurotoxicity by reducing glutamate levels in hippocampus. Esculetin, 7-benzyloxycoumarin, aryl or thioester coumarin confers MAO inhibitory activity and are significant antidepressant and anti-Parkinson's drug moieties.

Aromatic ring of coumarins have been proved to be efficient due to its anticholinesterase action along with suppression of amyloid-beta aggregation and beta secretase (Srikrishna D et al., 2016). Coumarin derivatives like coumarinyl semicarbazones, coumarinyl triazoles, coumarinyl oxadiazoles, and 7 substituted coumarins exhibited good epileptic activity in different screening models of epilepsy (1 Keri RS et al., 2022). Benzothiazine coumarins has anti-nociceptive property mediated by opioid receptors (Park SH et al. 2013).

Antimicrobial Action

Presence of a heterocyclic ring (imidazole, piperazine, fluoro-isoquinoline, azole, and thiazole) and long-chain derivatives of coumarin (ostruthin), anthogenol, imperatorin, aegelinol and agasyllin) are some of the coumarin analogs with potent anti-bacterial activity against Gram-positive and Gram negative bacteria. In addition, aminocoumarin analogs such as novobiocin, chlorobiocin, coumermycin and simocyclinone act by inhibiting bacterial DNA gyrase. Novobiocin also has anticancer action by inhibition of heat-shock proteins. To add on studies reveal that coumarins 4 acetic acid benzylidine hydrazide, coumarinyl thiazepines, 3,4 disubstituted coumarins possess anti tuberculosis effect (Srikrishna D et al., 2016, Cheke RS et al., 2022).

Further there are numerous evidences showing coumarins inhibitory role against various virus infection such as HIV, influenza, hepatitis, enterovirus and coxsackie virus by interfering with viral entry, replication or regulation of cellular pathways. Coumarin-linked benzoxazole-5-carboxylic acids showed the inhibitory activity against hepatitis virus. Bis-coumarin derivatives, coumarinyl triazines, tetra coumarins, coumarins metal complexes inophyllums, imperatorin and calanolides are reported to exhibit anti-HIV activity by inhibiting reverse transcriptase. Anti-antifungal coumarin derivative, such as psoralen, imperatorin, osthole and ostruthin are effective against fungal strains like Botrytis cinerea, Fusarium graminearum, Phytophthora capsici, Rhizoctonia solani, and Sclerotinia sclerotiorum. Ability of coumarin to activate immune system propelled it to be effective in treatment of other chronic infections such as brucellosis, mononucleosis, mycoplasmosis, toxoplasmosis and Q fever. Studies show that sulfonamide-attached coumarin and isocoumarins had anti-malarial potential akso (Mishra S et al., 2020).

Conclusion

Coumarins has become an important scaffold in the search for better therapeutic moieties due to its wide prevalence and broad spectrum of pharmacological action. Hence we cannot belittle the importance of this drug in the long run and there is no doubt that coumarins will etch its own place in the therapeutic scenario.

References

Bansal Y, Sethi P & Bansal G. Coumarin: a potential nucleus for anti-inflammatory molecules. *Med Chem Res*, 22, 3049-3060 (2013). https://doi.org/10.1007/s00044-012-0321-6.

Blahová J, Svobodová Z. Assessment of coumarin levels in ground cinnamon available in the Czech retail market. *The Scientific World Journal*, 2012; 2012:4. doi: 10.1100/2012/263851.263851.

Bourgaud F, Hehn A, Larbat R, Doerper S, Gontier E, Kellner S, Matern U. Biosynthesis of coumarins in plants: a major pathway still to be unravelled for cytochrome P450 enzymes. *Phytochemistry Reviews*, 2006;5(2-3):293-308. doi: 10.1007/s11101-006-9040-2.

Bubols GB, Vianna Dda R, Medina-Remon A, von Poser G, Lamuela-Raventos RM, Eifler-Lima VL, Garcia SC. The antioxidant activity of coumarins and flavonoids. *Mini Rev Med Chem,* 2013 Mar; 13(3):318-34. doi: 10.2174/138955713804999775. PMID: 22876957.3.

Campos-Toimil M, Orallo F, Santana L, Uriarte E. Synthesis and Vasorelaxant Activity of New Coumarin and Furocoumarin Derivatives, *Bioorganic & Medicinal Chemistry Letters,* 2002; 12, (5):783-786.

Cheke RS, Patel HM, Patil VM, Ansari IA, Ambhore JP, Shinde SD, Kadri A, Snoussi M, Adnan M, Kharkar PS, Pasupuleti VR, Deshmukh PK. Molecular Insights into Coumarin Analogues as Antimicrobial Agents: Recent Developments in Drug Discovery. *Antibiotics*, 2022, 11, 566. https://doi.org/10.3390/antibiotics11050566.

Menezes JCJMDS, Diederich MF. Natural dimers of coumarin, chalcones, and resveratrol and the link between structure and pharmacology. *European Journal of Medicinal Chemistry*, 2019; 182:p. 111637. doi: 10.1016/j.ejmech.2019.111637.

Kadhum AA, Al-Amiery AA, Musa AY, Mohamad AB. The antioxidant activity of new coumarin derivatives. *International Journal of Molecular Sciences*, 2011; 12(9):5747-5761. doi: 10.3390/ijms12095747.

Keri RS, Budagumpi S, Balappa Somappa S. Synthetic and natural coumarins as potent anticonvulsant agents: A review with structure-activity relationship. *J Clin Pharm Ther*, 2022 Jul; 47(7):915-931. doi: 10.1111/jcpt.13644. Epub 2022 Mar 15. PMID: 35288962.

Leal LK, Ferreira AA, Bezerra GA, Matos FJ, Viana S. Antinociceptive, anti-inflammatory and bronchodilator activities of Brazilian medicinal plants containing coumarin: A comparative study. *Journal of Ethnopharmacology*, 2000; 70(2):151-159. doi: 10.1016/S0378-8741(99)00165-8.

Li H, Yao Y, Li L. Coumarins as potential antidiabetic agents. *J Pharm Pharmacol*, 2017 Oct; 69(10):1253-1264. doi: 10.1111/jphp.12774. Epub 2017 Jul 3. PMID: 28675434.

Lu PH, Liao TH, Chen YH, Hsu YL, Kuo CY, Chan CC, Wang LK, Chern CY, Tsai FM. Coumarin Derivatives Inhibit ADP-Induced Platelet Activation and Aggregation. *Molecules*, 2022 Jun 23; 27(13):4054. doi: 10.3390/molecules27134054. PMID: 35807298; PMCID: PMC9268609.

Luo K, Jahufer M Z, Wu F, Di H, Zhang D, Meng X, Zhang J and Wang Y. Genotypic variation in a breeding population of yellow sweet clover (*Melilotus officinalis*) Frontiers in Plant. *Science*, 2016; 7:p. 972.

Martins Borges MF, Fernandes Roleira FM, Milhazes NJ, Villares EU, Penin LS. Simple coumarins: Privileged scaffolds in medicinal chemistry. *Frontiers in Medicinal Chemistry*, 2009; 4:23-85. Doi: 10.2174/978160805207310904010023.

Mishra S, Pandey A, Manvati S. Coumarin: An emerging antiviral agent. *Heliyon*, 2020 Jan 27; 6(1):e03217. doi: 10.1016/j.heliyon.2020.e03217. PMID: 32042967; PMCID: PMC7002824.

Park SH, Sim YB, Kang YJ, Kim SS, Kim CH, Kim SJ, Lim SM, Suh HW. Antinociceptive profiles and mechanisms of orally administered coumarin in mice. *Biol Pharm Bull*, 2013; 36(6):925-30. doi: 10.1248/bpb.b12-00905. PMID: 23727914.

Poonam V, Raunak GK, Reddy CSL, Kumar G, Jain R, Sharma KS, Prasad KA, Parmar SV. Chemical constituents of the genus Prunus and their medicinal properties. *Current Medicinal Chemistry*, 2011; 18(25):3758-3824. doi: 10.2174/092986711803414386.

Razavi BM, Arasteh E, Imenshahidi M, Iranshahi M. Antihypertensive effect of auraptene, a monoterpene coumarin from the genus Citrus, upon chronic administration. *Iran J Basic Med Sci*, 2015 Feb; 18(2):153-8. PMID: 25810889; PMCID: PMC4366726.

Sarker SD, Nahar L. Progress in the chemistry of naturally occurring coumarins. *Progress in the Chemistry of Organic Natural Products*, 2017; 106:241-304. doi: 10.1007/978-3-319-59542-9_3.

Sharanabasappa B. Patil K, Joshi H. Medicinal significance of novel coumarin analogs: Recent studies. *Journal of Applied Pharmaceutical Science,* 2012; 02 (06): 236-40.

Sharifi-Rad J, Cruz-Martins N, López-Jornet P, Lopez EP, Harun N, Yeskaliyeva B, Beyatli A, Sytar O, Shaheen S, Sharopov F, Taheri Y, Docea AO, Calina D, Cho WC. Natural Coumarins: Exploring the Pharmacological Complexity and Underlying Molecular Mechanisms. *Oxid Med Cell Longev,* 2021 Aug 23; 2021:6492346. doi: 10.1155/2021/6492346. PMID: 34531939; PMCID: PMC8440074.

Sharopov F, Setzer WN. Medicinal plants of Tajikistan. In: Egamberdieva D, Öztürk M, editors. Vegetation of Central Asia and environs. Switzerland: Springer. *Nature,* 2018. pp. 163-210.

Shimizu B. 2-Oxoglutarate-dependent dioxygenases in the biosynthesis of simple coumarins. *Frontiers in Plant Science,* 2014; 5:p. 549. doi: 10.3389/fpls.2014.00549.

Srikrishna D, Godugu C, Dubey PK. A Review on Pharmacological Properties of Coumarins. *2016 Mini Reviews in Medicinal Chemistry,* 16(999):1-1. DOI: 10.2174/1389557516666160801094919.

Tava A. Coumarin-containing grass: Volatiles from sweet vernalgrass (*Anthoxanthum odoratum* L.) *Journal of Essential Oil Research,* 2001; 13(5):367–370. doi: 10.1080/10412905.2001.9712236.

Wu Y, Xu J, Liu Y, Zeng Y and Wu G. A Review on Anti-Tumor Mechanisms of Coumarins. *Front Oncol,* 2020 10:592853. doi: 10.3389/fonc.2020.592853.

Chapter 6

An Overview of Coumarins: Pharmacognosy, Phytochemistry and Structural Activity Relationship (SAR)

Gnanakumar Prakash Yoganandam[1,*], PhD, Ramesh Vimalavathini[1], PhD and Meenatchisundaram Sakthiganapathi[2]

[1]Department of Pharmacognosy, College of Pharmacy,
Mother Theresa Post Graduate and Research Institute of Health Sciences (MTPG&RIHS),
A Government of Puducherry Institution, Puducherry, India
[2]Department of Pharmacognosy, School of Pharmacy,
Sri Balaji Vidyapeeth (SBV) (Deemed to be University), Puducherry, India

Abstract

Coumarins (2*H*-1-benzopyran-2-one) are named based on the plant *Coumarouna odorata* (*Dipteryx odorata*, Family: Fabaceae), from which it was first isolated by Vogel in 1820. Coumarins are secondary metabolites existing in a wide array of higher plants and also in some microorganisms and animal species. In the plant kingdom, Coumarin occurs in both monocotyledonous and dicotyledonous plants and found in plant families such as *Umbelliferae, Rutaceae, Compositae, Leguminosae, Oleaceae, Moraceae* and *Thymelaeacea*. Coumarins are present in different plant body part including roots, leaves, flowers, fruits, roots and in the exudates of plants. The biosynthesis of Coumarin in plants occurs through hydroxylation, glycolysis and cyclization of Cinnamic acid. Dietary exposure to benzopyrones is significant as these compounds are found in vegetables, fruits, seeds, nuts, coffee, tea, and wine. More than 1300 Coumarins have been identified as secondary metabolites from plants, bacteria, and fungi. The prototypical compound is known as 1, 2-benzopyrone or, less commonly, as hydroxycinnamic acid and lactone, and it has been well studied. Although distributed

---

* Corresponding Author's Email:gprakashyoga@gmail.com.

throughout all parts of the plant, the Coumarins occur at the highest levels in the fruits, Bael fruits (*Aegle marmelos*), seeds, Tonka beans (*Calophyllum inophyllum*), roots (*Ferulago campestris*), leaves (*Murraya paniculata*) and latex of the tropical rainforest tree (*Calophyllum teysmannii* var. *inophylloide*). They are also found at high levels in some essential oils such as Cassia oil, Cinnamon bark oil, and Lavender oil. Environmental conditions and seasonal changes could influence the incidence of Coumarins in varied parts of the plant. Structure activity relationship (SAR) attempts to establish the suitability of various functional groups or moieties at different positions of a pharmacophore nucleus and is thus exploited for optimization of drug receptor interaction. SAR studies of coumarins reveal that its electronegativity and aromaticity impart plethora of pharmacological activity.

Keywords: pharmacognosy, phytochemistry, SAR, coumarins, *Coumarouna odorata*, leguminosae, benzopyrone

Introduction

Coumarin and its derivatives are an important group of natural compounds widely distributed in the natural kingdom (Keating and O'Kennedy, 1997). They can be found in the integument of roots, stems, leaves, flowers, fruits, and seeds, although the highest concentration is generally in fruits and flowers (Miranda and Cuellar, 2001). Originally, coumarin was first isolated from the seed of *D. odorata*. Coumarins are secondary metabolites of higher plants, few microorganisms (bacteria and fungi), and sponges (De Lira et al., 2007). The function of this type of end product of secondary metabolism is related to defence mechanisms against herbivores and attacks by microorganisms. These compounds are biosynthesized from phenylalanine via the shikimic acid pathway (Dewick, 2002). Natural coumarins are generally unsaturated lactones and comprise another class of C_6-C_3 compounds. Almost all the natural coumarins have an oxygenated substituent at position-C_7, either free as in hydroxylated umbelliferone, or combined (methyl, sugars, etc.) in other derivatives. Structurally, they are measured derivatives of the ortho-hydroxy-cinnamic acid (Cai et al., 2006).

There are different groupings for the coumarin derivatives. Generally, they can be chemically classified according to the most common cores: simple, complex and miscellaneous coumarins. More complex coumarins are generally fused with other heterocycles (Venugopala et al., 2013). Therefore, they can be classified as: simple coumarins, furanocoumarins, dihydrofuranocoumarins, pyranocoumarins (linear and angular), phenylcoumarins, and biscoumarins (Borges et al., 2005). As said before, hundreds of coumarins have been identified in natural sources, especially plants. Major coumarin constituents isolated from plants include: simple hydroxycoumarins, furanocoumarins and isofuranocoumarins, pyranocoumarins, biscoumarins, and dihydroisocoumarins (Figure 1) (Borges et al., 2009).

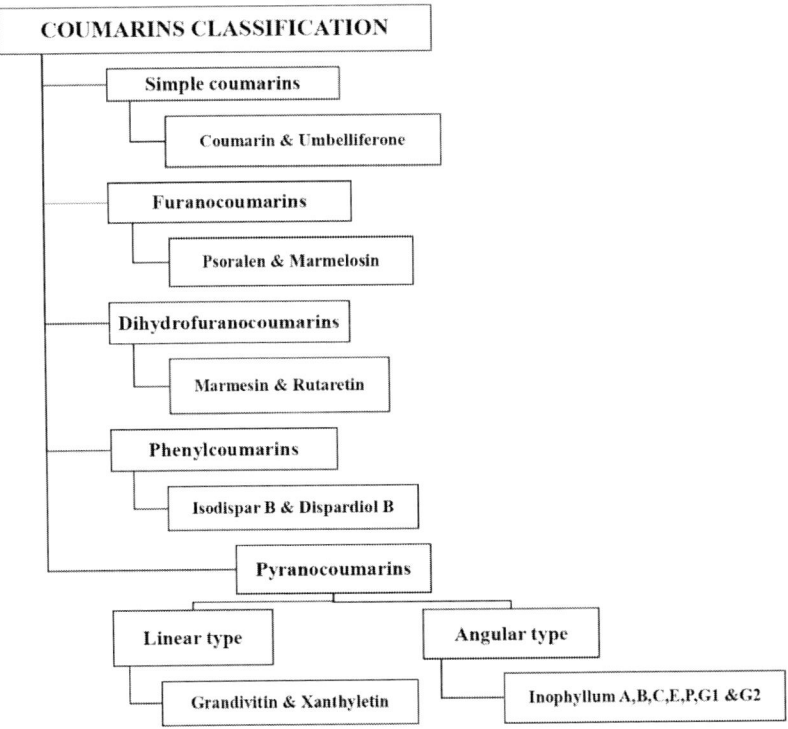

Figure 1. Principal types of coumarins isolated from plants.

Coumarins have been isolated from hundreds of plants species distributed in more than 40 different families. There were isolated more than different 1300 coumarins, well distributed in Angiospermae, Monocotyledoneae and Dicotyledoneae families. Orders with occurrence numbers >100 are Araliales, Rutales, Asterales, Fabales, Oleales, Urticales, and Thymelaeales. Families with occurrence numbers >100 are Apiaceae (Umbelliferae), Rutaceae, Asteraceae (Compositae), Fabaceae (Leguminosae), Oleaceae, Moraceae, and Thymelaeaceae, respectively (Figure 2) (Ribeiro and Kaplan, 2002). The best known and researched coumarins in the field of phytochemistry, pharmacology, medicinal chemistry, and the food science can be found in these families.

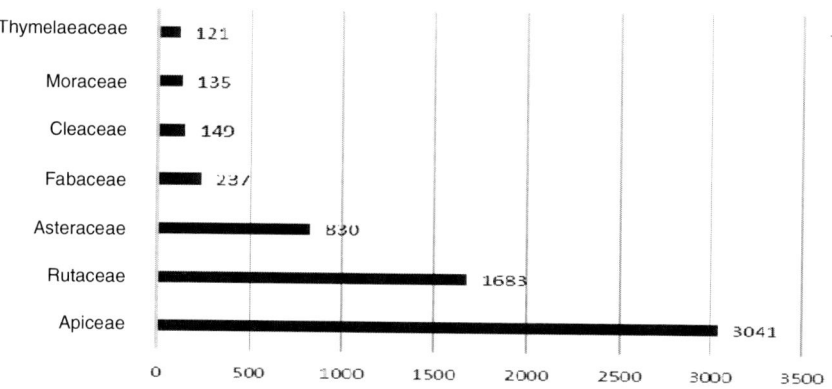

Figure 2. Number of coumarins presented in seven different families of plants.

Among the coumarins classified as "various" is the dicoumarol, which is formed by bacterial fermentation of Yellow Sweet Clover, and was isolated for the first time from decomposed leaves of *Melilotus albus* (Fabaceae/Leguminosae). An approximation for the dicoumarol biogenesis is the hydroxylation of the 4-position of the coumarin, that then captures a molecule of formaldehyde and is condensed with another molecule of 4-hydroxycoumarin, and finally enolize the keto group forming the dicoumarol (Dewick, 2002).

Table 1. List of major plant families containing Coumarins

S.No	Name of the plant family	Plant species	Common names	Name of the Coumarin
1	Umbelliferae or Apiaceae	*Ammi visnaga*	Toothpick weed	Marmesin, Xanthotoxin, Isopimpinellin
		Ammi majus	*Bishop's flower*	Imperatorin, bergapten,
		Anethum graveolens	Dill	Aesculetin, Bergapten, Scopoletin
		Coriandrum sativum	Coriander	Umbelliferone
		Cuminum cyminum	Cumin	Escopoloetina, Bergapten
		Daucus carota	Wild carot	8-methoxy psoralen, 5-methoxy psoralen
		Ferula foetida	Asafoetida	Asacoumarin A and Asacoumarin B
		Pimpinella anisum	Aniseed	Scopoletin, Umbelliferone, Umbelliprenine, Bergapten
		Trachyspermum ammi	Ajowan	Imperatorin, Isoimperatorin, Isopimpinellin,
2	Rutaceae	*Aegle marmelos*	Bael fruit	Sesqui, di and tri terpenic coumarin ethers,
		Citrus aurantium	Bitter Orange	Auraptenol, Bergaptol, Escoparonal.
		Murraya paniculata	Orange Jessamine	Coumarins
		Paramygnya monophylla	---------	Poncitrin, Nordentatin
		Toddalia aculeata	Orange climber	Ulopterol
		Zanthoxilum americanum	Northern Prickly Ash	Xanthyletin, Xanthoxyletin, Alloxanthoxyletin
3	Asteraceae/Compositae	*Ageratum conyzoides*	Mexican ageratum	1-2 benzopirone
		Cichorium intybus	Chicory	Coumarins
		Hieracium pilosela	Mouse Ear	Coumarins (0.2–0.6%): 7-glucosil-umbeliferone
		Matricaria recutita	Chamomille	Umbelliferone and its methyl ether, Heniarin.
		Lactuca virosa	Wild Lactuce	Aesculin, Cichoriin
4	Fabaceae/Leguminosae	*Dipteryx odorata*	Tonka Bean	Coumarins
		Glycyrriza glabra	Liquorice	Glycyrin, Heniarin, Liqcoumarin, Umbelliferone
5	Moraceae	*Morus alba*	White Mullberry	Coumarins

Table 1. (Continued)

S.No	Name of the plant family	Plant species	Common names	Name of the Coumarin
6	Oleaceae	*Oleae europaea*	Olive	Coumarins
7	Thymelaeaceae	*Daphne gnidium*	Flax-leaved daphne	Daphnetin, Daphnin, Acetyl umbelliferone, Daphnoretin
8	Achanthaceae	*Justicia pectoralis*	Tilo	Umbelliferone
9	Araliaceae	*Eleutherococcus senticosus*	Eleutherococcus	Coumarins
10	Brassicacae/Cruciferae	*Radicula armoracia*	Horse radish root	Aesculetin

Coumarins in Medicinal Plants

A large number of valuable species used commonly as medicinal plants, aromatic plants, and edible plants for human and animal feeding belongs to coumarin-rich plant families. Among them are species with well-documented biological activity, in which coumarins are part of the active principles. Table.1 shows a selection of plants of these families (first listed seven families with number of occurrence >100) and some other families with species of particular pharmacological interest on chronic diseases. Coumarins presenting great pharmacological interest have been isolated in different geographical regions from other botanical families. Also shown are the coumarin compounds having species and their yield (if available). Most of these plants are well known by people and scientists as part of herbal medicine repertories in Europe, Asia, or the Americas (Correa, 1984; Lemmens and Bunyapraphastara 2003; Newall et al., 1996; Peris et al., 1995). From the list, several coumarin-containing species or genera have also ethnomedical records in Cuba and the Caribbean Basin (Fuentes and Exposito, 1995; Robineau et al., 1996). Among of plant included are species with a great historical record of ethnomedicinal uses, and are present in traditional medicine systems: Ayurveda Medicine, Traditional Chinese Medicine and Unani Medicine, or in other recent cultures. Also, renowned species used on conventional therapeutics and modern herbal medicine are included, ie. *Aesculus hippocastanum* (Horse-chest nut), *Passiflora incarnata* (Passion Flower), *Lawsonia inermis* (Henna), *Hypericum perforatum* (Saint John Wort), *Tilia cordata* (Lime Tree) and *Uncaria tomentosa* (Cat's Claw).

Coumarins are also present in several species belonging to different botanical families, which are widespread in the northeastern region of Brazil (Leal et al., 2000). Some of them are reported in folk medicine as traditional remedies drugs for the treatment of respiratory diseases. It is the great structural diversity of coumarinic compounds that allows for their several applications, and also allows for the high interest of these derivatives as phytochemicals. The pharmacological and biochemical properties and therapeutic applications of simple coumarins depend upon the pattern of substitution (Kostova and Mojzis, 2007).

SAR of Coumarin

Structure activity relationship (SAR) attempts to establish the suitability of various functional groups or moieties at different positions of a pharmacophore nucleus and is thus exploited for optimization of drug receptor interaction. The coumarin structure is derived from cinnamic acid via ortho-hydroxylation, trans-cis isomerisation and lactonisation. The cis form seems to be unstable and isomerises to Trans form. SAR studies of coumarins reveals that its electronegativity and aromaticity imparts plethora of pharmacological activity. The pharmacological activities will vary depending upon the position and substitution pattern on the coumarin ring. Alternation of structure of coumarin has been executed by modification of coumarin scaffold at 3, 4, 5, 6, 7 and 8 position, (Figure 3) hybridization with other analogs or incorporation of a metal ion into coumarin derivatives. This confers advantages such as enhancement of pharmacological action or yielding hybrids with multiple activities. For example coumarin nucleus with lipoic acid confers both the antioxidant and anti-inflammatory properties. An increase in activity of such of coumarin metal complexes has been reported along with promising antioxidant, anticancer or antibacterial activities (Balewski L. et al., 2021). However unsubstituted coumarins are toxic because they undergo oxidative decarboxylation causing byproducts which form complexes with heavy metals in human body (Srikrishna et al., 2016; Jain et al., 2012).

SAR for Anti-Inflammatory Action of Coumarin

Research activities reflect that synthetic coumarin derivatives such as coumarin-3- carboxamides, 4-aryl coumarins, coumarinyl benzimidazole, 7-alkyloxy coumarins and acyl dihydroxy coumarin derivatives exhibit synergistic antioxidant and anti-inflammatory properties. Natural coumarins like umbelliferone, scopoletin, columbiatnetin, visniadin, marmin, prenyloxycouarins and prenyloxyfuranocoumarins also exhibit potent anti-inflammatory as well as antioxidant activities (Srikrishna et al., 2016; Bubols et al., 2013). In general, substitution with nitrogen containing heterocyclic group at C-3, unsaturated heterocyclic rings at C-4, chlorine or bromine at C-6, inclusion of carbamates and nitrogen-containing heterocyclic group at C-7 and carbonyl or hydroxyl group at the C-8 position imparts potent anti-inflammatory activity. But both in natural and synthetic coumarins C-5 position remains unexplored for anti-inflammatory activity (Bansal et al., 2013; Grover et al., 2015).

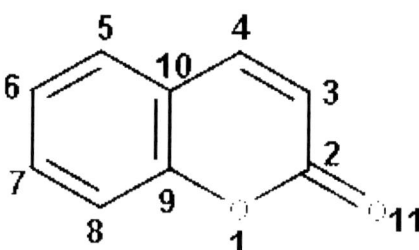

Figure 3. Chemical structure of Coumarin (2 H- chromen-2-one).

SAR for Antiepileptic Action of Coumarins

Literature evidences shows that the coumarin analog substituted at all position with varied substituents yields potent anticonvulsant compounds. Coumarin derivatives like coumarinyl semicarbazones, coumarinyl triazoles, coumarinyl oxadiazoles, and 7 substituted coumarins exhibited good epileptic activity in different screening models of epilepsy. For anticonvulsant action the 1st and 2nd position of coumarin has to be unsubstituted. Presence of heterocyclic groups like thiazole, oxazole, oxadiazole, triazole and thiazolidinone rings at 3rd position, halo -methyl, oxadiazole, imidazolium and benzimidazolium groups in 4th position. thiourea, iminothiazoline and thiazoline-5- carboxylic acid derivatives at 5th position, alkylthioureas and thiazole at 6th position, 4-chloroaryl group at 7th position and aryl group connected to 1,2,4-oxadiazole ring, hydroxyl and alkene group at 8th position of coumarin scaffold imparts excellent antiepileptic activity. Natural coumarins containing oxyprenyl or prenyl along with C-5 side chain such as phellopterin, imperatorin, isoimperatorin, oxypeucedanin, osthole and heraclenin enhanced gamma amino butyric acid (GABA) activity thus imparting anti-epileptic activity (Srikrishna et al., 2016; Keri et al., 2022).

SAR for Anti-Neurodegenerative Properties Coumarins

Esculetin, 7-benzyloxycoumarin, aryl or thioester coumarin confers monoamine oxidase (MAO) inhibitory activity and hence are significant antidepressant and anti-Parkinson's drug moieties. Substitution on 3-arylcoumarin skeleton with methyl, methoxy, nitro, amino groups and bromine atoms at *meta-* and *para-*positions of this group along with hydroxyl or o-alkyl group at 7th position is crucial for anti-MAO activity. Studies reveal that aromatic ring of coumarins imparts anticholinesterase action along with inhibition of beta secretase and amyloid-beta accumulation (Srikrishna et al., 2016, Sharifi-Rad et al., 2021).

SAR for Anticoagulant Properties Coumarins

Studies show that 4-aminocoumarin and 4-hydroxy coumarins derivatives inhibit coagulation and platelet aggregation and are promising anticoagulant coumarins. Minimum requirement for anticoagulant activity of coumarin are a 4-hydroxy group, 3-substituent and a bis molecule (Srikrishna et al., 2016, Jain et al., 2012).

SAR for Anticancer Activity of Coumarins

Coumarin derivatives with thiophene, furan and isatin at 3rd position along with the hydroxyl or o-aryl or o-alkyl group at 4th 7th and 8th position confer potent anticancer properties (Srikrishna et al., 2016).

SAR for Metabolic Modulation by Coumarins

Coumarin hybrids such as bis-indole coumarin with ethyl ester and coumarin-dihydro quinazolinone analogs exhibited profound anti-hyperlipidemic activity. The position and number of free hydroxyl groups in coumarin-resveratrol hybrid seems to play a vital part in deciding the tyrosine kinase inhibitory activity of coumarin. Incorporation of C8/C7-C6 and C6/C5-C6 in 4-flavonoid-coumarin analog improved α-glucosidase inhibition and glucose consumption thus imparting better antidiabetic activity than their flavone and coumarin alone lead structures (Gupta et al., 2019).

SAR for Antimicrobial Actions of Coumarins

Presence of a heterocyclic ring (imidazole, piperazine, fluoro-isoquinoline, azole, and thiazole), long-chain derivatives of coumarin (ostruthin, anthogenol, imperatorin, aegelinol and agasyllin) and aminocoumarin analogs (novobiocin chlorobiocin, coumermycin and simocyclinone) are some of the important coumarin analogs with potent anti-bacterial activity against Gram-positive and Gram negative bacteria. SAR requirement for antibacterial activity includes presence of thiazole and pyrimidine rings at 3rd position, substitution with halo, methyl groups at 6th position and hydroxyl group at 4th and 7th position of coumarin. To add on studies reveal that coumarin benzylidine hydrazide, coumarinyl thiazepines, 3,4 disubstituted coumarins possess anti tuberculosis effect.

Coumarin-linked benzoxazole-5-carboxylic acids showed anti-hepatitis virus activity. Bis-coumarin derivatives, coumarinyl triazines, tetra coumarins, coumarins metal complexes inophyllums, imperatorin and calanolides are reported to exhibit anti-HIV activity by inhibiting reverse transcriptase. Anti-antifungal coumarin derivative, such as psoralen, imperatorin, osthole and ostruthin are effective against various strains of fungi. Studies show that sulfonamide-attached coumarin, coumarin-triazole, coumarin-pyrazoline hybrids and isocoumarins had profound anti-malarial potential (Cheke et al., 2022).

Conclusion

Coumarin and its types are playing vital role as an important group of natural compounds extensively spread in the natural kingdom. It is the great structural diversity of coumarinic compounds that allows for their several applications, and also allows for the high interest of these derivatives as phytochemicals. The pharmacological and biochemical properties and therapeutic applications of simple coumarins depend upon the pattern of substitution Coumarins and its analogs obtained from either natural or synthetic sources exhibited vital biological potential and further analysis of SAR helps in the discovery, synthesis and development of compounds with more potency and multiple modes of action.

Declarations

None.

References

Balewski L., Szulta S., Jalińska A., Kornicka A. A Mini-Review: Recent Advances in Coumarin-Metal Complexes with Biological Properties. *Front Chem.* (2021) 1(9):781779. doi: https://doi.org/10.3389/fchem.2021.781779.

Bansal Y., Sethi P., Bansal G. Coumarin: a potential nucleus for anti-inflammatory molecules. *Med Chem Res* (2013); 22, 3049–3060. doi: https://doi.org/10.1007/s00044-012-0321-6.

Borges F., Roleira F., Milhazes N., Santana L., Uriarte E. Simple coumarins and analogues in medicinal chemistry: occurrence, synthesis and biological activity. *Curr Med Chem.* 2005;12(8):887-916. doi: https://doi.org/10.2174/0929867053507315.

Borges F., Roleira F., Milhazes N., Uriarte E., Santana L. Simple coumarins: Privileged scaffolds in medicinal chemistry. *Front Med Chem Biol Inter.* 2009;4:23-85. doi: https://doi.org/10.2174/978160805207310904010023.

Bubols G. B., Vianna Dda R., Medina-Remon A., von Poser G., Lamuela-Raventos R. M., Eifler-Lima V. L., Garcia S. C. The antioxidant activity of coumarins and flavonoids. *Mini Rev Med Chem.* (2013);13(3):318-34. doi: https://doi.org/10.2174/138955713804999775.

Cai Y., Sun M., Xing J., Luo Q., Corke H. Structure-radical scavenging activity relationships of phenolic compounds from traditional Chinese medicinal plants. *Life Sci.* 2006;78:2872-88. doi: https://doi.org/10.1016/j.lfs.2005.11.004.

Cheke R. S., Patel H. M., Patil V. M., Ansari I. A., Ambhore J. P., Shinde S. D., Adel Kadri, Mejdi Snoussi, Mohd Adnan, Prashant S. Kharkar, Visweswara Rao Pasupuleti, Prashant K. Deshmukh. Molecular Insights into Coumarin Analogues as Antimicrobial Agents: Recent Developments in Drug Discovery. *Antibiotics* 2022, 11, 566. doi: https://doi.org/10.3390/antibiotics11050566.

Correa P. M. Diciona´rio de plantas u´ teis do Brasil e das exoticas cultivadas. Brazil: Ministério da Agricultura, Instituto Brasileiro de Desenvolvimento, Florestal; 1984 (*Dictionary of useful plants from Brazil and cultivated exotics.* Brazil: Ministry of Agriculture, Brazilian Institute of Development, Forestry; 1984).

De Lira S. P., Seleghim M. H. R., Williams D. E., Marion F., Hamill P., Jean F., Raymond J. Andersen, Eduardo Hajdu, Roberto G. S. Berlinck. A SARS-coronovirus 3CL protease inhibitor isolated from the marine sponge Axinella cf. corrugata: structure elucidation and synthesis. *J Braz Chem Soc.* 2007;18 (2). doi: https://doi.org/10.1590/S0103-50532007000200030.

Dewick P. M. *Medicinal Natural Products: A Biosynthetic Approach.* 2a ed. England: John Wiley & Sons Ltd; 2002. doi: https://doi.org/10.1002/0470846275.

Fuentes V., Exposito A. Las enecuestas botanicas sobre plantas medicinales en Cuba. Revista del Jardin Botanico Nacional. 1995;26:77-145 (Botanical surveys on medicinal plants in Cuba. *Magazine of the National Botanical Garden.* 1995;26:77-145).

Grover J. and Jachak S. M. *Coumarins as privileged scaffold for anti-inflammatory drug development RSC Adv,* (2015); 5, 38892.

Gupta M., Kumar S., Chaudhary S. Coumarins: a unique scaffold with versatile biological behaviour. *Asian J Pharm Clin Res* (2019); 12(3): 27-38.

Jain P., Joshi H. Coumarin: Chemical and Pharmacological Profile. *Journal of Applied Pharmaceutical Science* (2012); 02 (06):236-240.

Keating G. J., O'Kennedy R. *The chemistry and occurrence of coumarins.* O'Kennedy RTRD, editor. England: John Wiley & Sons West Sussex; 1997.

Keri RS, Budagumpi S, Balappa Somappa S. Synthetic and natural coumarins as potent anticonvulsant agents: A review with structure-activity relationship. *J Clin Pharm Ther.* (2022);47(7):915-931. doi: https://doi.org/10.1111/jcpt.13644.

Kostova I, Mojzis J. Biologically active coumarins as inhibitors of HIV-1 (RT, IN and PR). *Fut HIV Ther.* 2007;1(3):315-29. doi: https://doi.org/10.2217/17469600.1.3.315.

Leal L. K. A. M., Ferreira A. A. G., Bezerra G. A., Matos F. J. A., Viana G. S. B. Anticonceptive, anti-inflammatory and bronchodilator activities of Brazilian medicinal plants containing coumarin: a comparative study. *Journal of Ethnopharmacology.* 2000;70 (2):151-9. doi: https://doi.org/10.1016/S0378-8741(99)00165-8.

Lemmens R. H. M. J., Bunyapraphastara N. Plant resourses of South-East Asia. In: Lemmens RHMJ, Bunyapraphastara N, editors. *Medicinal and poisonous plant.* Leiden (Holanda): Backhuys; 2003.

Miranda M., Cuellar A. Farmacognosia y productos naturales. La Habana: Félix Varela; 2001 (*Pharmacognosy and natural products.* Havana: Felix Varela; 2001).

Newall C., Anderson L., Phillipson J. Herbal medicines. *Aguide for health-care professionals*. London: The pharmaceutical Press; 1996.

Peris J. B., Stübing G., Vanaclocha B. Fitoterapia aplicada. edicion r, editor. Valencia MICOF; 1995 (*Vanaclocha B. Applied herbal medicine*).

Ribeiro C. V., Kaplan M. A. Tendências evolutivas de famílias produtoras de cumarinas em angiospermae. Quim Nova. 2002;25(4):533-8 (Evolutionary trends of coumarin-producing families in angiosperms. *Kim Nova*. 2002;25(4):533-8).

Sharifi-Rad J., Cruz-Martins N., López-Jornet P., Lopez E. P., Harun N., Yeskaliyeva B., Beyatli A., Sytar O., Shaheen S., Sharopov F., Taheri Y., Docea A. O., Calina D., Cho W. C. Natural Coumarins: Exploring the Pharmacological Complexity and Underlying Molecular Mechanisms. *Oxid Med Cell Longev*. (2021) 23;:6492346. doi: https://doi.org/10.1155/2021/6492346.

Srikrishna D., Godugu C., Dubey P. K. A Review on Pharmacological Properties of Coumarins. *Mini Reviews in Medicinal Chemistry* (2016); 16(999):1-1. doi: https://doi.org/10.2174/1389557516666160801094919.

Venugopala K. N., Rashmi V., Odhav B. Review on Natural Coumarin Lead Compounds for Their Pharmacological Activity. *BioMed Research International*. 2013;2013:1-14. doi: https://doi.org/10.1155/2013/963248.

Index

A

Alzheimer's disease (AD), 78
animals, viii, 23, 25, 49, 95
antibacterial activity, 54, 56, 116
anticancer action, 99, 102
anticancer activity, 42, 49, 50, 52, 53, 64, 99, 100, 115
anticoagulant, 26, 43, 101, 115
application, viii, 24, 26, 27, 30, 37, 43, 62, 66, 68, 70, 74

B

benzopyrone, x, 20, 95, 107, 108
bioactivities, viii, 23, 24, 26, 29, 42, 43, 44, 49, 60, 62
biosynthesis of coumarins, 47, 49
bromocoumarins, v, vii, 1, 2, 5, 10

C

C-3 position, 42, 50, 53, 57
C-4 position, 29, 35, 39, 43, 53, 73
C-6 position, 28, 44
C-7 position, 25, 28, 29, 33, 45, 46, 53
C-8 position, 28, 29, 46, 49, 114
cancer, vii, ix, 11, 13, 20, 24, 25, 26, 42, 43, 49, 50, 51, 52, 53, 62, 67, 68, 69, 72, 73, 74, 93, 98, 99, 100
catechol-O-methyl transferase (COMT), 78
cellulose sulfuric acid (CSA), v, vii, 1, 2, 3, 4, 5, 6, 9, 10, 11
chemistry, vii, 1, 13, 20, 22, 62, 64, 65, 66, 67, 68, 69, 70, 71, 72, 73, 74, 79, 91, 94, 95, 103, 104, 105, 110, 117, 118, 119
coumarin, iii, v, vii, viii, ix, 2, 4, 13, 14, 15, 16, 17, 20, 21, 22, 24, 25, 26, 27, 28, 29, 32, 34, 35, 36, 37, 39, 40, 41, 42, 43, 44, 45, 46, 47, 48, 49, 50, 51, 52, 53, 54, 55, 56, 57, 58, 59, 60, 61, 62, 63, 64, 65, 66, 67, 68, 69, 70, 71, 72, 73, 74, 75, 77, 78, 80, 83, 85, 86, 87, 88, 89, 91, 93, 94, 95, 96, 97, 98, 99, 100, 101, 102, 103, 104, 105, 107, 108, 109, 110, 111, 112, 113, 114, 115, 116, 117, 118, 119
coumarin significance, 13
coumarins, v, vii, viii, ix, 1, 2, 13, 14, 15, 16, 20, 21, 22, 23, 24, 25, 26, 27, 28, 29, 30, 31, 32, 33, 34, 35, 36, 37, 38, 39, 40, 41, 42, 43, 44, 45, 46, 47, 49, 50, 51, 53, 54, 55, 57, 58, 60, 61, 62, 63, 64, 65, 66, 67, 68, 69, 71, 72, 73, 75, 77, 79, 80, 88, 93, 94, 95, 96, 98, 99, 100, 101, 102, 103, 104, 105, 107, 108, 109, 110, 111, 112, 113, 114, 115, 116, 117, 118, 119
Coumarouna odorata, ix, 93, 94, 107, 108

D

diabetes, 60, 99, 101
distribution, 2, 78, 94, 98

E

electrophilic substitution reactions, 4
extraction, 30, 31, 32, 36

F

furancoumarin(s), 28

K

Knoevenagel synthesis, 37, 40

L

lactone, ix, x, 28, 31, 34, 35, 93, 95, 107

Index

leguminosae, viii, ix, 23, 25, 107, 108, 110, 111

M

MAO-A, 78, 81, 82, 87, 89, 91
MAO-B, v, vii, viii, 77, 78, 80, 81, 82, 83, 84, 85, 86, 87, 88, 89, 91, 92, 102
medicinal applications, 13
metal-free, 2, 3
molecular docking, v, vii, ix, 65, 77, 78, 80, 83, 85, 89
monoamine oxidase, viii, 77, 78, 90, 91, 102, 115
monoamine oxidase (MAO), v, vii, viii, 72, 77, 78, 79, 80, 81, 82, 83, 84, 85, 86, 87, 88, 89, 90, 91, 92, 102, 115
multidrug resistance (MDR), 25, 46, 100

N

neurodegenerative disease, viii, 77, 78, 99, 102
neurodegenerative diseases, viii, 77, 78, 99, 102
neurodegenerative disorder(s), 78

O

occurrence, ix, 94, 110, 112, 117, 118
one-pot procedure, 2, 10
one-pot synthesis, v, vii, 1, 2, 3, 5, 6, 11, 68
one-pot synthesis of bromocoumarins, vii, 1, 3, 6, 11

P

Parkinson's disease (PD), 78
Pechmann condensation, v, vii, 1, 2, 3, 4, 21, 38, 65, 71

Perkin synthesis, 37
pharmacognosy, v, 93, 107, 108, 118
phytochemistry, v, vii, ix, 68, 69, 72, 93, 94, 103, 107, 108, 110
plants, vii, ix, 2, 14, 20, 23, 25, 29, 44, 47, 49, 52, 62, 67, 93, 94, 95, 97, 103, 104, 105, 107, 108, 109, 110, 112, 117, 118
pyranocoumarin(s), 29, 67

R

reactive oxygen species (ROS), 26, 63, 78, 99
Remier-Tiemalni synthesis, 41

S

SAR, v, x, 64, 72, 80, 87, 88, 107, 108, 113, 114, 115, 116, 117
separation, 30, 31, 32
simple coumarins, viii, 23, 27, 28, 38, 48, 105, 109, 113, 117
sodium chloroacetate synthesis, 41
structure-activity relationship, v, vii, 46, 53, 54, 62, 77, 78, 87, 104, 118
substitution reactions, 36
synthesis, viii, 2, 5, 10, 11, 16, 17, 20, 21, 22, 24, 33, 36, 37, 38, 39, 40, 41, 42, 43, 44, 45, 47, 49, 50, 52, 62, 63, 64, 65, 66, 67, 68, 69, 70, 71, 72, 73, 74, 77, 99, 103, 117, 118
synthetic approaches, 13, 16

V

Vilsmeier-Haack synthesis, 42

W

Wittig synthesis, 37, 39, 40, 73